EXAM PREPARATION
Firefighter I & II

EXAM PREPARATION
Firefighter I & II

Andrea Walter
Marty Rutledge
Chris Hawley

THOMSON

DELMAR LEARNING ™

Australia Canada Mexico Singapore Spain United Kingdom United States

Exam Preparation for Firefighter I & II

Andrea Walter, Marty Rutledge, Chris Hawley

Vice President, Technology and Trades SBU:
Dave Garza

Director of Learning Solutions:
Sandy Clark

Acquisitions Editor:
Alison Weintraub

Product Manager:
Jennifer A. Thompson

Channel Manager:
William Lawrensen

Marketing Coordinator:
Mark Pierro

Production Director:
Mary Ellen Black

Senior Production Manager:
Larry Main

Senior Production Editor:
Thomas Stover

Technology Project Specialist:
Linda Verde

Editorial Assistant:
Maria Conto

ISBN: 1-4018-9923-4

NOTICE TO THE READER

Contents

Preface

Congratulations to you as you embark on your career in the fire service! Countless hours of work and training have led you to this point. The questions within these pages, as well as in the CD in the back of this book, will help prepare you for the firefighter certification exam.

How To Use This Book

Based on the 2002 Edition of NFPA 1001 for firefighter training and *Firefighter's Handbook: Essentials of Firefighting and Emergency Response, Second Edition*, each exam was created to ensure that all firefighter competencies at levels I and II were covered. As such, each exam contains questions from the chapters in the book, as well as references to applicable NFPA Standard requirements. In addition, because the questions within this book are also designed based on the NFPA Standard, the book will prepare you for any firefighter I and II exam.

The book is organized in a logical sequence based on Bloom's Taxonomy of Learning, that progresses from test-taking strategies to questions about memorizing important terms and on to concepts and application of those concepts in a given scenario:

- The best place to start is at the introductory chapter, ***Acing Your Certification Exam.*** Here you will find valuable test-taking strategies, as well as advice on how to set up an effective study schedule prior to the certification exam date.
- ***Phase One: Knowledge and Comprehension*** contains three practice exams of 100 questions each to gauge your knowledge and ability to retain important facts and concepts.
- ***Phase Two: Application and Analysis*** contains three practice exams of 100 questions each that evaluate your ability to solve problems and recognize how important concepts fit together.
- ***Phase Three: Evaluation and Synthesis*** is the most complex stage, also containing three practice exams of 100 questions each, and judges how accurately you can make decisions based on given scenarios and your knowledge of the subject matter.
- ***Phase Four: Final Exam*** is a 200–question test combining questions from all three phases, simulating the certification exam.
- ***Back of Book CD:*** contains *all* exams in a self-grading ExamView format, allowing you the option of self-study. Also included is a bonus final exam of 200 questions for additional practice in computer-based testing.
- ***Answers to Questions, NFPA and Textbook References and Rationale*** are provided for each question in all of the exams, allowing you to track your progress as well as the applicable sources for additional study as needed.

Successful candidates will take all of these exams in the order provided, using the answers, references, and rationale to grade their own work. The CD provides an excellent opportunity to retake tests as needed, as well as a bonus final exam. A copy of *Firefighter's Handbook: Essentials of Firefighting and Emergency Response, Second Edition*, the reference material for the practice tests is also a useful tool for further study.

(Order #: 1-4018-3575–9)

Features of this Book

This book provides many features to enhance your learning experience and help lead you to closer to your goal of successful completion of the certification exam:

- *Test Taking Strategies* and *Study Guidelines* are outlined in the introductory chapter, allowing you to effectively practice, and fully prepare, for the exam.
- *Questions of increasing complexity* are organized in a logical sequence of the book to ensure accomplishment of important concepts related to the officer role.
- *Answers to the Questions* allow you to track your progress.
- *NFPA References* correlate the questions to the applicable NFPA 1001 Standard requirements illustrating the competency which the question covers.
- *Textbook References* correlate the questions to the appropriate pages in the *Firefighter's Handbook, Second Edition* book, to allow for further study if needed.
- *Rationale* also accompanies each question, providing you with the necessary explanation to support the correct answer.
- *Questions in ExamView format* on the CD in the back of the book, allow you to use this book as a self-study tool and to practice computer-based testing. A bonus final exam is also provided.

About the Authors

Mr. Hawley is a Project Manager in the National Security Programs division of Computer Sciences Corporation (CSC), Alexandria, Virginia. Mr. Hawley supports the Department of Defense International Counterproliferation program, (ICP) in course development, delivery, and client support. The ICP program provides several WMD courses throughout Eastern Europe, Central Asia and other parts of the world. Prior to his current position, Mr. Hawley served as a Fire Specialist and Special Operations Coordinator for the Baltimore County Fire Department in Towson, Maryland. In this capacity, he served as the coordinator for the Hazardous Materials Response and Advanced Technical Rescue Teams. He has 24 years experience as a firefighter and 17 years as a HazMat responder. Mr. Hawley is a published author of five texts on hazardous materials and terrorism response. He has written numerous magazine and trade journal articles, and also assisted many publishers in the review and development of other emergency services texts and publications.

Marty Rutledge is a Firefighter/Engineer and EMS Program Manager for Loveland Fire and Rescue in Loveland, Colorado.Marty is a member of the Fire Certification and Advisory Board to the Colorado Division of Fire Safety, serving also as the State First Responder program coordinator. He is also a member of the Colorado State Fire Fighter's Association and has over 14 years of fire and emergency services experience in both volunteer and career ranks.Marty also authored and served as technical expert for a supplementary firefighter training package for Delmar Learning's *Firefighter's Handbook, Second Edition.*

Andrea A. Walter is a Firefighter/Technician with the Metropolitan Washington Airports Authority and a life member and former officer of the Sterling Volunteer Rescue Squad. Ms. Walter has been active in the fire and emergency services community for many years, serving as the Manager of the Commission on Fire Accreditation International for the International Association of Fire Chiefs and assisting in a variety of projects with the National Volunteer Fire Council, Women in the Fire Service, and the United States Fire Administration. She has more than 16 years of experience in the fire and emergency services. In addition to being an author for this text, Ms. Walter is also an author and the Content Editor for Delmar Learning's *Firefighter's Handbook, Second Edition* and *First Responder Handbook.*

About the Series Advisor

Mike Finney is the head of the department of public safety services at Great Oaks Institute of Technology and Career Development. He has served on the NFPA 1041 committee in past years and is a current board member for the International Association of Fire Service Instructors. His area of expertise is teaching and curriculum development.

Acing the Certification Exam: An Introduction to Test-Taking Strategies

Introduction

Test time. Whether you are preparing for a certification test or a hiring test, the thought of an examination strikes fear in many people's hearts. The fear is so common that psychologists even have a diagnosis called test anxiety. However, testing does not have to be that way. Evaluations are simply an instrument to determine if you were effectively taught the information intended, or if you have the knowledge base necessary to do the job. That's all! If the purpose of testing is so simple, why do so many people become so anxious when test time comes. Several factors play into test anxiety and why so many people have such fears of testing. However, these can be overcome. With the assistance of this guide, you too can be better prepared and calmer on examination day.

Differences between hiring and certification tests

A hiring test is intended to determine if an applicant has the necessary knowledge base to perform the job. Different departments have different requirements for the job that is being filled, and those doing the hiring want to know that the applicant can meet the challenge. The outcome is very simple—get the most qualified applicant for the position. This can be the opportunity for you to "show your stuff." Hiring tests are designed to give each applicant an objective avenue to present their skill base. Careful preparation will give you the opportunity to excel. Every applicant has the same test, same questions, and same opportunity to prepare. As an applicant, take this opportunity to show the skills and knowledge base you have. Success on the hiring test is not necessarily measured on a set score or a "passing" score. Some departments set a cut-off score for applicants to proceed to the next phase, but many will take the top scores for the next phase. In this situation, it is imperative to get the highest score when compared to your competitors.

A certification test takes on a different dimension. Certification tests are designed to demonstrate mastery and are measured to national consensus standards. Since the other students do not measure the success, there is typically a set score that you must reach to be successful. The minimum score is set rather than being driven by all people taking the test.

How are test questions developed?

How questions are developed is critical to the understanding of how to take a test. Whether the examination is a hiring test, a promotional test, or a certification examination, they are developed around a clear set of objectives.

The learning objectives are a very important piece of the educational process. The objectives tell you several things:

- Under what conditions the student should be able to apply the knowledge. (condition)
- What type of knowledge the student should gain. (behavior)
- What depth or level of understanding the student should have. (standard)

While the order may vary based on the preference of the developer, this condition, behavior, and standard approach should be addressed in all learning objectives. When reviewing the learning objectives, take a moment to break it down into the individual pieces. This will provide insight into how the class should be taught. One will gain a wealth of information as to the intent of the designer by studying the learning objectives.

Domains of learning

Based on Bloom's Taxonomy, there are three primary areas or domains around which testing is designed. Knowing the domains will give an indication of the evaluation approach. The three domains of learning are:

- Cognitive domain- primarily deals with intellectual or knowledge skills, those that are intellectual processing or mental learning.
- Psychomotor domain- primarily deals with physical skills, those that require physically doing a task.
- Affective domain- primarily deals with a mindset, those that require a change in attitude or behavior.

Within the three domains were levels from the most basic understanding to advanced. Since we are primarily focusing on written tests, we will only deal with the cognitive domain. Table I-1 gives you the breakdown of the levels of understanding for the cognitive domain. The key to remember is that the higher the level you reach, the greater the understanding of the material must exist.

Table I-1

Knowledge: Recall of data	Examples: Recite a policy. Quote prices from memory to a customer. Knows the safety rules. Keywords: defines, describes, identifies, knows, labels, lists, matches, names, outlines, recalls, recognizes, reproduces, selects, states
Comprehension: Understand the meaning, translation, interpolation, and interpretation of instructions and problems. State a problem in one's own words.	Examples: Rewrites the principles of test writing. Explain in one's own words the steps for performing a complex task. Translates an equation into a computer spreadsheet. Keywords: comprehends, converts, defends, distinguishes, estimates, explains, extends, generalizes, gives examples, infers, interprets, paraphrases, predicts, rewrites, summarizes, translates
Application: Use a concept in a new situation or unprompted use of an abstraction. Applies what was learned in the classroom into novel situations in the workplace.	Examples: Use a manual to calculate an employee's vacation time. Apply laws of statistics to evaluate the reliability of a written test. Keywords: applies, changes, computes, constructs, demonstrates, discovers, manipulates, modifies, operates, predicts, prepares, produces, relates, shows, solves, uses
Analysis: Separates material or concepts into component parts so that its organizational structure may be understood. Distinguishes between facts and inferences.	Examples: Troubleshoot a piece of equipment by using logical deduction. Recognize logical fallacies in reasoning. Gathers information from a department and selects the required tasks for training. Keywords: analyzes, breaks down, compares, contrasts, diagrams, deconstructs, differentiates, discriminates, distinguishes, identifies, illustrates, infers, outlines, relates, selects, separates

Synthesis: Builds a structure or pattern from diverse elements. Put parts together to form a whole, with emphasis on creating a new meaning or structure.	Examples: Write a company operations or process manual. Design a machine to perform a specific task. Integrates training from several sources to solve a problem. Revises and processes to improve the outcome. Keywords: categorizes, combines, compiles, composes, creates, devises, designs, explains, generates, modifies, organizes, plans, rearranges, reconstructs, relates, reorganizes, revises, rewrites, summarizes, tells, writes
Evaluation: Make judgments about the value of ideas or materials.	Examples: Select the most effective solution. Hire the most qualified candidate. Explain and justify a new budget. Keywords: appraises, compares, concludes, contrasts, criticizes, critiques, defends, describes, discriminates, evaluates, explains, interprets, justifies, relates, summarizes, supports

Cognitive Domain (courtesy of http://www.nwlink.com/~donclark/hrd/bloom.html)

NFPA standards

The movement toward national standards for the fire service began in 1971 with the Joint Council of National Fire Service Organizations. The intent was to develop national performance standards. The end result, today, is 67 levels for 16 standards dealing with the professional qualifications for the fire service outlined by standards developed by the NFPA. Committees made up of training leaders, educators, private industry, and technical specialists guide standards development and revision. The intent is to give a better standard that meets a greater need. The standards are reviewed and revised as needed every five years. (This is why it is important when referencing an NFPA standard to also reference the year.) As a result, performance standards are kept up-to-date and accurate to what is needed in the field. While NFPA standards are not federal mandates, they do provide a widely accepted standard to follow. They provide the most current expectations for the position referenced.

The NFPA standards also provide an excellent reference for developing job descriptions. The 1000- series are referred to as the Professional Qualifications standards. Each provides a solid outline of performance requirements for almost any given position with the fire service. They are also easily adaptable to most any department. While there are other standards that can be followed, those set forth by the NFPA are the most widely recognized and are considered foundational for any department.

How does this apply to me?

NFPA standards are important to someone taking a fire-related examination, because often they are the basis for the learning objectives for test development. For certification examinations, NFPA standards are the foundation for many certifying bodies. For hiring and promotional tests, they are a solid basis for creating objective tests. A good resource to begin preparing for certification and hiring tests is the appropriate NFPA standard. Review the specific level and the requirements for such a level. This will give you some indication of what the test developer will be considering. As well, look at the requisite knowledge and skills. This will give you an indication of the extent of understanding of the information. An excellent way of reviewing the standards is to look at the verbs of the job performance requirement (JPR) in the context of Bloom's Taxonomy. Much like the requisite information, the JPR, as it relates to Bloom's Taxonomy, will tell you the level of understanding you must have with the information.

Test obstacles

Test obstacles are issues that complicate test taking. If we view test taking as simply an avenue to determine the individual's comprehension of the material, then test obstacles are barriers to the process. There are many issues that may create test obstacles. We will discuss a few.

Mental

Mental test obstacles can sometimes be the greatest hurdles to overcome. Mental preparation for a test can be as important as intellectual preparation. So often, many people have failed an exam before they even begin. Issues that arise out of mental obstacles are:

- feeling unprepared
- feeling incompetent
- fear of taking tests
- fear of failure

Overcoming these obstacles can be your greatest asset when testing. Not allowing yourself to be beaten before entering the testing area, can make the difference in success and failure on the exam.

Physical

Improper rest, poor eating habits, and lack of exercise can be some of the physical obstacles to overcome. When preparing for tests, always ensure that you get plenty of rest the night before, have a well-balanced meal before the test, and ensure you have a regiment of proper exercise. Physical obstacles are typically the easiest to overcome, however, the most overlooked.

Emotional

The emotional obstacles are often the most vague with which to deal. Much like the mental, emotional obstacles can cause a person to do poorly on an exam well before they enter the room. Stress-related issues that can interfere with test taking are:

- family concerns
- work-related concerns
- financial concerns

Emotional issues can cause a person to lose focus, cloud decision-making skills, and become a distracter. Overcoming these obstacles requires a conscious effort to ensure that the emotional does not interfere with the test.

Preparing to Take a Test

BEFORE the Test

1. Start preparing for the examination. For certification exams, start the first day of class. You can do this by reading your syllabus carefully to find out when your exams will be, how many there will be, and how much they are weighed into your grade. For hiring exams, it is recommended to begin studying at least eight weeks before the test.

2. For certification classes, plan reviews as part of your regular weekly study schedule; a significant amount of time should be used to review the entire material of the class.

3. Reviews are much more than reading and reviewing class assignments. You need to read over your class notes and ask yourself questions on the material you don't know well. (If your notes are relatively complete and well organized, you may find that very little rereading of the textbook for detail is needed.) You may want to create a study group for these reviews to reinforce your learning.

4. Review for several short periods rather than one long period. You will find that you are able retain information better and get less fatigued.

5. Turn the main points of each topic or heading into questions and check to see if the answers come to you quickly and correctly. Do not try to guess the types of questions; instead concentrate on understanding the material.

DURING the Test

1. Preview the test before you answer anything. This gets you thinking about the material. Make sure to note the point value of each question. This will give you some ideas on how best to allocate your time.

2. Quickly calculate how much time you should allow for each question. A general rule of thumb is that you should be able to answer 50 questions per hour. This averages out to one question every 1.2 seconds. However, make sure you clearly understand the amount of time you have to complete the test.

3. Read the directions CAREFULLY. (Can more than one answer be correct? Are you penalized for guessing? etc.) Never assume that you know what the directions say.

4. Answer the easy questions first. This will give you the confidence and give you a feel for the flow of the test. Only answer the ones for which you are sure of the correct answer.

5. Go back to the difficult questions. The questions you have answered so far may provide some indication of the answers.

6. Answer all questions (unless you are penalized for wrong answers).

7. Generally, once the test begins, the proctor can ONLY reread the question. He/she cannot provide any further information.

8. Circle key words in difficult questions. This will force you to focus on the central point.

9. Narrow your options on the question to two answers. Many times, a question will be worded with two answers that are obviously inaccurate, and two answers being close. (However, only one is correct.) If you can narrow your options to two, guessing may be easier. For example, if you have four options on a question, then you have a 25% chance of getting the question correct when guessing. If you can narrow the options to two answers, then you increase to a 50% chance of getting the correct choice.

10. Use all of the time allotted for the test. If you have extra time, review your answers for accuracy. However, be careful of making changes on questions of which you are not sure. Often times, people change the answer of questions of which they were not sure, when their first guess was correct.

AFTER the Test

Relax. The test has been turned in. You can spend hours second-guessing what you "could" have done, but the test is complete. For certification tests, follow up to see if you can find out what objectives you did well and what areas you could improve. Review your test if you can; otherwise, try to remap the areas of question and refocus your studying.

Preparation Plan

Once you have acquired the reference texts for the examination, begin by reviewing the introduction, the table of contents, and review how the book is organized. The introduction will tell you how the book has been set up and how it is intended to assist the individual with the learning process. The table of contents will provide a snapshot of how the book is organized. Scanning the text will give you an overview of the book's design. Once you have done this, break the chapters into four review sections. Using Table I-2 on the next page, fill in week 8 with the first section, week 7 with the second section, week 6 with the third section, and week 5 with the fourth section. Focus your energy into 50-minute increments with 10-minute breaks. Consider one hour for each chapter. However, on chapters in which you are competent, less time can be spent than with chapters that are more unfamiliar.

Week 4 will be spent taking section one of the Exam Prep guide. Base your time on 50 questions per hour. (100 questions should take 2 hours.) Do not check the answers until you have completed the entire test. Any questions missed should be reviewed and ensure you have an understanding of why the answer is correct.

Week 3 will be spent taking section two of the Exam Prep guide. Again, base your time on 50 questions per hour. At the end of the test, check answers and correct wrong answers.

Week 2 should be spent on section three of the Exam Prep guide. Maintain timeframes and check answers at the end of the test.

Week 1 should be spent with section four of the Exam Prep guide.

Five days before, go through the section one tests again.

Four days before, go through the section two tests again.

Three days before, go through the section three tests again.

Two days before, go through the section four tests again.

One day before do a light review of the text, focusing on areas you missed on the practice tests. However, take the evening and relax. Do something you enjoy, but make sure it is not a late night. Go to bed early and make sure you get a good night's sleep.

The day of the test make sure you have a well-balanced breakfast and arrive at the test site early.

Table I-2

PREPARATION GUIDE *Plan based on a two-month schedule**	STUDY NOTES
Week 8 – Reference Text, Section I	
Week 7 –Reference Text, Section II	
Week 6 – Reference Text, Section III	
Week 5 – Reference Text, Section IV	
Week 4 – Exam Prep, Section I • Exam One • Exam Two • Exam Three	
Week 3 – Exam Prep, Section II • Exam One • Exam Two • Exam Three	
Week 2 – Exam Prep, Section III • Exam One • Exam Two • Exam Three	
Week 1 – Exam Prep, Section IV • Final Exam • Bonus Final Exam (on CD)	
5 days before – Review Section I Tests	

4 days before – Review Section II Tests	
3 days before - Review Section III Tests	
2 days before – Review Section IV Tests	
1 day before – Light review • Relax • Go to bed early	
Day of Test • Good breakfast • Arrive early	

Summary

Test taking does not have to be overwhelming. The obstacles to testing can be overcome and conquered through solid strategies and preparation. Initiating an effective plan, following it, and mentally preparing for a test can be your greatest tools to test success. As you work through the sections of this book, use the time as well, to work through some of the obstacles you face. When taking the tests in each section, try to simulate the environment of the actual test as much as possible. Successful testing is not an art. It is a learned skill. Through planning and practice anyone can acquire these skills.

KNOWLEDGE & COMPREHENSION

Section one is designed to evaluate your basic understanding of the material. In this section, we are testing you understanding of definitions, recalling information, and identifying terms. Referring to Table I-1 (Bloom's Taxonomy, Cognitive Domain), we are covering the following levels:

- knowledge
- comprehension

Having mastered section one, you should be able to have a basic comprehension of the material.

1. The testing of Class B extinguishers and agents uses a wood cribbing test.
   a. True
   b. False

2. Respiratory protection provided by SCBA is necessary, even during exterior defensive operations.
   a. True
   b. False

3. Proximity PPE is not designed for fire entry.
   a. True
   b. False

4. A mission statement is a written declaration by a fire agency describing the things that it intends to do to protect its citizenry or customers.
   a. True
   b. False

5. Structural PPE is dangerous to firefighters working at a wildland fire.
   a. True
   b. False

6. A solid tip is a type of fog nozzle.
   a. True
   b. False

7. Head pressure measures the pressure at the top of a column of water.
   a. True
   b. False

8. Protection systems are also known as auxiliary appliances.
   a. True
   b. False

9. A British thermal unit (Btu) is a measurement of heat that describes the amount of heat required to raise 1 pound of water by 10F.
   a. True
   b. False

10. Booster hose is smaller in diameter than hose used for structural firefighting operations.
    a. True
    b. False

11. The leading cause of firefighter deaths is overexertion/stress, which includes heart attacks as a result of stress.
    a. True
    b. False

12. A fire shelter is considered a first-line of defense against wildland fires.
    a. True
    b. False

13. Firefighting is a team effort. Working alone or outside the action plan endangers individuals and the team.
    a. True
    b. False

14. Protective systems have a major drawback in that they require a human to detect the fire and manually activate an alarm.
    a. True
    b. False

15. When selecting a fire extinguisher, it is very important that a firefighter be able to classify the fire into one of 5 classes.
    a. True
    b. False

16. Rescue operations such as confined space, collapse, rope, and trench always require full structural PPE.
    a. True
    b. False

17. Emergency communications centers should be equipped with backup power supplies.
    a. True
    b. False

18. Two national organizations provide certification of state and local firefighter training programs in accordance with NFPA standards: the National Board of Fire Service Professional Qualifications and the International Fire Service Accreditation Congress.
    a. True
    b. False

19. A thermal imaging camera or an electronic heat sensor is a useful tool in salvage operations.
    a. True
    b. False

20. When performing a fire safety inspection, the company officer should insist that the crew be escorted during the inspection.
    a. True
    b. False

21. A(n) _____ is the smallest particle into which an element or a compound can be divided without changing its chemical and physical properties.

    a. molecule

    b. atom

    c. oxidizer

    d. neutron

22. The most common type of lug to assist in making or breaking a connection is the _____.

    a. rocker lug

    b. spanner wrench

    c. Storz

    d. Higbee

23. The pressure, which for Earth is 14.7 pounds per square inch at sea level, is called _____ pressure.

    a. residual

    b. static

    c. atmospheric

    d. flow

24. The four elements of the fire tetrahedron are fuel, heat, _____,and chemical reaction.

    a. oxidation

    b. oxygen

    c. combustion

    d. organic

25. A _____ has automatic sprinklers attached to pipes with water under pressure all the time.

    a. retarded chamber

    b. wet pipe sprinkler system

    c. hydro advance sprinkler system

    d. OS&Y sprinkler system

26. Seat-mounted SCBA _____.

    a. allows a firefighter to don the SCBA en route to an emergency

    b. allows for less cylinder failures due to the storage technique

    c. interferes with bunker gear donning en route to a call

    d. can interfere with the cylinder gauge check

27. The accident chain includes the following components: the environment, _____, equipment, the event, and the injury.
    a. personnel policies
    b. procedures
    c. human factors
    d. mitigation

28. What incident command designation is given to a set of responders who are responsible for all operations within an assigned geographic area?
    a. A task force
    b. A branch
    c. A group
    d. A division

29. A team of firefighters with apparatus assigned to perform a specific function in a designated response area is a _____.
    a. section
    b. company
    c. crew
    d. group

30. NFPA 1581 requires that clothing be cleaned _____ as a minimum.
    a. once a week
    b. once a month
    c. yearly
    d. every six months

31. _____ will cause a PASS device to send a reminder to the wearer.
    a. Nothing
    b. Wearer inactivity
    c. A change in cardiac rhythm
    d. Lack of breathing

32. Lumber arranged in a systematic stack to support an unstable load is _____.
    a. shoring
    b. cribbing
    c. tunneling
    d. padding

33. _____ hose is used by trained personnel to fight fires.
    a. Attack
    b. Booster
    c. Supply
    d. Soft suction

34. _____ is simply a flow of electrons from a place where there are electrons to a place where electrons are lacking.
    a. Oxidation
    b. Electricity
    c. Nuclear energy
    d. Friction

35. A Class B fire involves _____.
    a. ordinary combustibles
    b. flammable and combustible liquids, gases, and greases
    c. combustible metals and alloys
    d. energized electrical equipment

36. In addition to voice communications, _____ 9-1-1 service provides emergency communications centers with the telephone number and address of the phone from which the call is originating.
    a. enabled
    b. basic
    c. encoded
    d. enhanced

37. Which WMD materials are closely related to pesticides?
    a. Explosives
    b. Nuclear
    c. Blister agents
    d. Nerve agents

38. Sprinkler orientation refers to _____.
    a. nothing (This is not a term used in protective systems.)
    b. male or female sprinkler head
    c. sprinkler geographic location on the line
    d. up, down, or sideways mounted

39. In the fire tetrahedron, _____ is considered the reducing agent in the chemical reaction of fire.
    a. heat
    b. oxygen
    c. fuel
    d. combustion

40. A red box is the label on a fire extinguisher for _____.
    a. ordinary combustibles
    b. flammable and combustible liquids, gases, and greases
    c. energized electrical equipment
    d. combustible metals and alloys

41. The simplestand often quickestway to stop water flow from an individual sprinkler head is to _____.
    a. insert a stop
    b. turn off the valve head
    c. reinsert a fusible link
    d. find the water main

42. A blue circle is the label for a fire extinguisher for _____.
    a. ordinary combustibles
    b. flammable and combustible liquids, gases, and greases
    c. energized electrical equipment
    d. combustible metals and alloys

43. A(n) _____ is used to connect a smaller hose to a larger one.
    a. reducer
    b. adapter
    c. increaser
    d. double male

44. Natural adverse conditions affecting fire streams do not include _____.
    a. wind and wind direction
    b. gravity
    c. air friction
    d. electricity

45. The _____ nozzle was designed for aircraft operations.
    a. fog
    b. piercing
    c. ARFF ram
    d. straight-bore

46. A flammable vapor fire or explosion requires that _____.
    a. temperature of the liquid be very low
    b. proper air-to-fuel mixture
    c. it be above the UEL
    d. it be below the LEL

47. A communications center may also be referred to as a PSAP, or public safety _____ point.
    a. awareness
    b. alarm
    c. arrival
    d. answering

48. Which category of chemical WMD agents may present with delayed symptoms?

    a. Blister agents

    b. Nerve agents

    c. Explosives

    d. Irritants

49. The _____ detector measures temperature increases above a predetermined rate.

    a. progressive heat

    b. fixed temperature

    c. thermal increase

    d. rate-of-rise

50. Nozzle reaction is _____.

    a. the movement of the nozzle opposite from the direction of the water flow

    b. the movement of the nozzle in the same direction as the water flow

    c. dangerous to persons weighing less than the nozzle operator

    d. inversely proportionate to the gpm

51. An important consideration of PPE is that it should be _____.

    a. clean and dry when worn

    b. kept away from UVA rays

    c. stored on hangers, rather than being folded

    d. replaced on a six-month basis

52. The pressure in the system with no hydrants or water flowing is _____ pressure.

    a. residual

    b. static

    c. atmospheric

    d. flow

53. _____ is/are a more predictable water source with less chance of contamination.

    a. Surface water

    b. Deep wells

    c. Natural springs

    d. Tanks, ponds, and cisterns

54. Ionization detectors use _____ to detect smoke.

    a. radioactivity

    b. light obscuration

    c. light scattering

    d. ionizes

55. In a tender shuttle operation, the portable tank is erected or inflated at the _____.
    a. hydro site
    b. water supply site
    c. drop-off point
    d. dump site

56. _____ are used in emergency communications centers to track the locations of emergency incidents and what units have been assigned to respond to emergency incidents.
    a. Mobile data computers
    b. Computer-aided dispatch systems
    c. Databases
    d. Command systems

57. The four elements of a fire stream are _____.
    a. pump, water, hose, and nozzle
    b. source, hose, pressure, and nozzle
    c. psi, gpm, length of hose, and friction loss
    d. type of source, diameter of hose, length of hose, and gallons per minute

58. An oxygen-deficient atmosphere is an atmosphere with oxygen concentrations below _____.
    a. 50%
    b. 100%
    c. 19.5%
    d. 21%

59. A Class D fires involves _____.
    a. ordinary combustibles
    b. flammable and combustible liquids, gases, and greases
    c. energized electrical equipment
    d. combustible metals and alloys

60. To tie ropes of unequal diameter together, a firefighter should use a _____.
    a. Becket bend
    b. half Becket bend
    c. triple Becket bend
    d. double Becket bend

61. Aluminum cylinders have a(n) _____ service life.
    a. 5-year
    b. unlimited
    c. 15-year
    d. limited

62. The most likely tools that a terrorist would use are_____.
    a. nerve agents
    b. nuclear device
    c. explosives
    d. anthrax

63. Fire hose is made in three types of construction. Which of these is not one of these types?
    a. Wrapped
    b. Plastic
    c. Braided
    d. Woven

64. Which is not a state of matter?
    a. Solid
    b. Melting point
    c. Liquid
    d. Gas

65. Drafting is simply _____.
    a. creating a vacuum inside the pump
    b. externally pumping source water into the pump
    c. using a jet siphon to force water into the pump from the dump site
    d. involves travel to and from the scene while using code three response

66. A(n) _____ is an organic compound containing only carbon and hydrogen.
    a. oxidizer
    b. molecule
    c. hydrocarbon
    d. proton

67. _____ is the measurable amount of pressure being exerted against a confined container by a liquid substance as it converts to a gas.
    a. Vapor density
    b. Evaporation
    c. Vapor pressure
    d. Diffusion

68. The minimum quantity on a truck that requires placarding for a Table 2 material is _____.
    a. 1,001 pounds
    b. 10,000 pounds
    c. 1 pound
    d. 1,000 liters

69. Hydraulics is the study of fluids _____.
    a. and their expansion capability
    b. and their effect on fire
    c. at rest and in motion
    d. none of the above

70. _____ is the process of minimizing the chance, degree, or probability of damage, loss, or injury.
    a. Risk management
    b. The safety chain
    c. The safety triad
    d. Vicarious experience

71. The communication process includes four basic elements: receive, understand, _____, and communicate.
    a. reference
    b. record
    c. read
    d. report

72. The temperature that a material would move from a solid to a liquid is what property?
    a. Contact
    b. Freezing point
    c. Melting point
    d. Gas point

73. What is the ideal method for transfer of command?
    a. By face-to-face meeting
    b. By telephone
    c. By radio
    d. By written report

74. A hydrant mounted on the wall of a building is called a _____ hydrant.
    a. wet barrel
    b. dry
    c. dry barrel
    d. wall

75. The combustion process can affect _____.
    a. long- and short-term health of firefighters
    b. short-term health of firefighters
    c. long-term health of firefighters
    d. firefighter morale

76. When using any nontrunked two-way radio, it is important to depress and hold the push-to-talk button at least _____ second(s) before talking to avoid clipping the first part of the message.

    a. one
    b. two
    c. three
    d. four

77. _____ devices are used when large volumes of water are required.

    a. Fog
    b. Master stream
    c. Combination
    d. Hand-held straight stream

78. Positive pressure self-contained breathing apparatus use pressure of _____, depending on the manufacturer.

    a. .5 to 1 psi
    b. 1 1/2 to 2 psi
    c. 10 psi
    d. 100 psi

79. Blunt end threads use a _____ to help firefighters align the Higbee cuts.

    a. Higbee cut
    b. Higbee thread
    c. Higbee ditch
    d. Higbee indicator

80. The difference between NFPA 1500 and OSHA regulations is _____.

    a. the AHJ must adopt the NFPA standard as policy
    b. NFPA supercedes OSHA
    c. whether or not the state is OSHA compliant
    d. a political matter better left to the company officers to decide

81. A(n) _____ on a sprinkler system allows the fire department pumper to supplement the water supply.

    a. gated Y
    b. fire department gate
    c. outside screw and yoke
    d. Siamese connection

82. The acronym PASS stands for _____.

    a. Pull, Aim, Squeeze, Sweep
    b. Point, Aim, Squeeze, Sweep
    c. Point, Aim, Sight, Sweep
    d. Pull, Aim, Sight, Sweep

83. The biggest problem with PASS devices results when wearers simply forget to turn their units on. This simple mental lapse has contributed to numerous firefighter fatalities. Because of this, the NFPA now requires _____.

    a. integrated PASS devices

    b. buddy checking

    c. PAR

    d. extra batteries

84. _____ that meets NFPA standards adds another layer of reasonable protection under wildland, proximity, and structural ensembles.

    a. Underwear

    b. Wildland gear

    c. A work uniform

    d. Polyblend t-shirts

85. A(n) _____ is a device that allows citizens to communicate with a telecommunicator through the use of a keyboard over telephone circuits instead of voice communications.

    a. CAD

    b. APCO

    c. TDD

    d. NENA

86. Mobile water supply apparatus, according to NFPA 1901, Standard for Automotive Fire Apparatus, must have a minimum of a _____ -gallon water tank.

    a. 500

    b. 750

    c. 1,000

    d. 1,500

87. A(n) _____ is a catalyst in the breakdown of molecules and possesses a chemical property that can pull apart a molecule and break apart the bond that previously existed.

    a. oxidizer

    b. proton

    c. electron

    d. hydrocarbon

88. The _____ is mandated by legislation to investigate all incident-related firefighter fatalities.

    a. National Fire Protection Association

    b. National Institute of Occupational Safety and Health

    c. Occupational Safety and Health Administration

    d. United States Fire Administration

89. A 2A extinguisher _____.
    a.   will extinguish twice the fire of a Class B replacement
    b.   is the smallest acceptable replacement for a 1BC extinguisher
    c.   is not adequate for fire service use
    d.   will extinguish twice the fire of a 1A extinguisher

90. Within an Incident Management System, there are four section chiefs: operations, logistics, _____, and finance/administration.
    a.   safety
    b.   staging
    c.   planning
    d.   command

91. Acids have a pH range of _____.
    a.   0-2
    b.   5-9
    c.   0-6.9
    d.   7.1-14

92. Which WMD materials use the term SLUDGEM to describe the exposure symptoms?
    a.   Blister agents
    b.   Nerve agents
    c.   Nuclear agents
    d.   Biological agents

93. Sublimation ability means that _____.
    a.   a material moves to a solid
    b.   a solid moves to the gaseous state
    c.   a liquid moves to the gaseous state
    d.   the material freezes

94. _____ describes the weight of a gas as compared to normal air and is identified as a number.
    a.   Vapor density
    b.   Evaporation
    c.   Vapor pressure
    d.   Diffusion

95. _____ are supervisory-level positions and are responsible for both firefighters and administrative duties.
    a.   Firefighters
    b.   Rescue specialists
    c.   Company officers
    d.   Driver/operators

96. The _____, which is a part of the Department of Labor, is responsible for the enforcement of safety-related regulations in the workplace.

    a. National Fire Protection Association

    b. Code of Federal Regulations

    c. Occupational Safety and Health Administration

    d. United States Fire Administration

97. A(n) _____ is a device that converts the entered code into paging codes, which then activate a variety of paging devices.

    a. CAD

    b. encoder

    c. home alerting device

    d. mobile data terminal

98. The NFPA standard entitled Standard for Fire Fighter Professional Qualifications is also known as _____.

    a. NFPA 1001

    b. NFPA 1500

    c. NFPA 472

    d. NFPA 1003

99. Stored hose _____.

    a. can last indefinitely

    b. should be used occasionally

    c. should occasionally be refolded

    d. can last for two years and then must be discarded

100. The legal term _____ means for every emergency medical incident, an emergency responder should treat the patient in the same manner as would another emergency responder with the same training.

    a. consent

    b. abandonment

    c. standard of care

    d. implied care

Phase I, Exam I: Answers to Questions

1. T	26. D	51. A	76. B
2. T	27. C	52. B	77. B
3. T	28. D	53. B	78. B
4. T	29. B	54. A	79. D
5. T	30. A	55. D	80. A
6. F	31. B	56. B	81. D
7. F	32. B	57. A	82. A
8. T	33. A	58. C	83. A
9. F	34. B	59. D	84. C
10. T	35. B	60. D	85. C
11. T	36. D	61. B	86. C
12. F	37. D	62. C	87. A
13. T	38. D	63. B	88. B
14. F	39. C	64. B	89. D
15. T	40. B	65. A	90. C
16. F	41. A	66. C	91. C
17. T	42. C	67. C	92. B
18. T	43. C	68. A	93. B
19. F	44. D	69. C	94. A
20. T	45. B	70. A	95. C
21. A	46. B	71. B	96. C
22. A	47. D	72. C	97. B
23. C	48. A	73. A	98. A
24. B	49. D	74. D	99. B
25. B	50. A	75. A	100. C

Phase I, Exam I:
Rationale & References for Questions

Question #1. The test for Class B extinguishers involves igniting a pan of flammable liquid (heptane), allowing a pre-burn period, and attacking the fire. NFPA 1001: 5.3.16. *FFHB, 2E:* Page 194.

Question #2. Respiratory protection provided by SCBA is necessary, even during exterior defensive operations. NFPA 1001: 5.3.1. *FFHB, 2E:* Page 143.

Question #3. It is important to note that proximity PPE is not designed for fire entry (to be totally enveloped by fire). NFPA 1001: 5.1.1.2. *FFHB, 2E:* Page 130.

Question #4. A mission statement is a written declaration by a fire agency describing the things that it intends to do to protect its citizenry or customers. NFPA 1001: 5.1.1.1. *FFHB, 2E:* Page 4.

Question #5. Fighting wildland fires with a structural ensemble can invite strained necks, heat stroke, and sprained ankles. NFPA 1001: 5.1.1.2. *FFHB, 2E:* Page 130.

Question #6. The two basic types of nozzles are solid stream (also called a smooth bore, straight bore, or solid tip) and fog nozzles with different styles available for each kind, especially fog nozzles. NFPA 1001: 5.3 NFPA 6.3. *FFHB, 2E:* Page 282.

Question #7. Head pressure measures the pressure at the bottom of a column of water in feet. Head pressure can be gained or lost when water is being pumped above or below the level of the pump. NFPA 1001: 5.3 NFPA 6.3. *FFHB, 2E:* Page 291.

Question #8. They are also called auxiliary appliances. NFPA 1001: 6.5.1. *FFHB, 2E:* Page 309.

Question #9. A British thermal unit (Btu) is a measurement of heat that describes the amount of heat required to raise 1 pound of water by 1°F. NFPA 1001: 5.3; 6.3. *FFHB, 2E:* Page 77.

Question #10. Booster hose is smaller diameter, rubber-coated hose of 3/4 or 1-inch size usually mounted on reel that can be used for outside fires or overhaul operations after the fire is out. NFPA 1001: 5.3.10. *FFHB, 2E:* Page 222.

Question #11. The leading cause of firefighter deaths is overexertion/stress, which includes heart attacks as a result of stress. NFPA 1001: 6.1.1.1. *FFHB, 2E:* Page 108.

Question #12. A fire shelter is a last-resort protective device for firefighters caught or trapped in an environment where a firestorm or blow-up is imminent. NFPA 1001: 5.1.1.2. *FFHB, 2E:* Page 132.

Question #13. Firefighting is a team effort. Working alone or outside the action plan endangers individuals and the team. NFPA 1001: 6.1.1.1. *FFHB, 2E:* Page 119.

Question #14. Detection systems are varied. Some require people to detect the fire and manually activate an alarm. Others are highly complex systems that can detect a fire almost at its ignition and sound an alarm. NFPA 1001: 6.5.1. *FFHB, 2E:* Page 309.

Question #15. An understanding of the fire classes leads to selection of the proper unit and agent. NFPA 1001: 5.3.16. *FFHB, 2E:* Page 185.

Question #16. Rescue operations such as confined space, collapse, rope, and trench do not necessarily require full structural PPE. . *FFHB, 2E:* Page 133.

Question #17. Emergency communications centers should be equipped with backup power supplies. NFPA 1001: 5.2.1. *FFHB, 2E:* Page 50.

Question #18. Two national organizations provide certification of state and local firefighter training programs in accordance with NFPA standards: the National Board of Fire Service Professional Qualifications and the International Fire Service Accreditation Congress. NFPA 1001: N/A. *FFHB, 2E:* Page 26.

Question #19. A thermal imaging camera or an electronic heat sensor is a useful tool in overhaul operations. NFPA 1001: 5.3.13. *FFHB, 2E:* Page 650.

Question #20. When performing a fire safety inspection, the company officer should insist that the crew be escorted during the inspection. The escort will be able to grant access to all areas of the building and answer any questions that may arise. NFPA 1001: 5.5.1. *FFHB, 2E:* Page 662.

Question #21. A molecule is the smallest particle into which an element or a compound can be divided without changing its chemical and physical properties. NFPA 1001: 5.3; 6.3. *FFHB, 2E:* Page 79.

Question #22. The most common type of lug is the rocker lug, but pins and recessed pins are also used. NFPA 1001: 5.1.1.1. *FFHB, 2E:* Page 226.

Question #23. Atmospheric pressure The pressure exerted by the atmosphere, which for Earth is 14.7 pounds per square inch at sea level NFPA 1001: 5.3.15. *FFHB, 2E:* Page 217.

Question #24. The four elements of the fire tetrahedron are fuel, heat, oxygen, and chemical reaction. NFPA 1001: 5.3; 6.3. *FFHB, 2E:* Page 77.

Question #25. A wet pipe sprinkler system has automatic sprinklers attached to pipes with water under pressure all the time. NFPA 1001: 6.5.1. *FFHB, 2E:* Page 318.

Question #26. From the seated position, it is extremely difficult, if not impossible, to check the cylinder gauge and compare it to the regulator gauge. NFPA 1001: 5.3.2. *FFHB, 2E:* Page 163.

Question #27. The accident chain includes the following components: the environment, human factors, equipment, the event, and the injury. NFPA 1001: 6.1.1.1. *FFHB, 2E:* Page 110.

Question #28. The incident command designation division is given to a set of responders who are responsible for all operations within an assigned geographic area. NFPA 1001: 6.1.1.1. *FFHB, 2E:* Page 41.

Question #29. A team of firefighters with apparatus assigned to perform a specific function in a designated response area is a company NFPA 1001: 5.1.1.1. *FFHB, 2E:* Page 28.

Question #30. NFPA 1581, Standard on Fire Department Infection Control Program, requires that clothing be cleaned every six months as a minimum. NFPA 1001: 5.1.1.2. *FFHB, 2E:* Page 136.

Question #31. Inactivity for thirty seconds causes the device to send a chirp or other reminder to the wearer. NFPA 1001: 5.1.1.2. *FFHB, 2E:* Page 135.

Question #32. As previously defined, cribbing is the use of various dimensions of lumber arranged in systematic stacks to support an unstable load. NFPA 1001: 6.4.1. *FFHB, 2E:* Page 496.

Question #33. Attack hose is used by trained personnel to fight fires. NFPA 1001: 5.3.15. *FFHB, 2E:* Page 222.

Question #34. Electricity is simply a flow of electrons from a place where there are electrons to a place where electrons are lacking. NFPA 1001: 5.3; 6.3. *FFHB, 2E:* Page 84.

Question #35. Class B fires involve flammable and combustible liquids, gases, and greases. NFPA 1001: 5.3.16. *FFHB, 2E:* Page 185.

Question #36. In addition to voice communications, enhanced 9-1-1 service provides emergency communications centers with the telephone number and address of the phone from which the call is originating. NFPA 1001: 5.2.1; 5.2.2. *FFHB, 2E:* Page 54.

Question #37. First responders should have a basic awareness of the threat of terrorism and basic response actions. NFPA 472: 4.2.1. *FFHB, 2E:* Page 927.

Question #38. Orientation means up, down, or sideways. NFPA 1001: 6.5.1. *FFHB, 2E:* Page 315.

Question #39. In the fire tetrahedron, fuel is considered the reducing agent in the chemical reaction of fire. NFPA 1001: 5.3; 6.3. *FFHB, 2E:* Page 82.

Question #40. Figure 8-3. NFPA 1001: 5.3.16. *FFHB, 2E:* Page 189.

Question #41. The simplest - and often quickest - way to stop water flow from an individual sprinkler head is to insert a stop. NFPA 1001: 6.5.1. *FFHB, 2E:* Page 326.

Question #42. Figure 8-3. NFPA 1001: 5.3.16. *FFHB, 2E:* Page 189.

Question #43. An increaser is used to connect a smaller hose to a larger one, and a reducer connects a larger hose to a smaller one. NFPA 1001: 5.1.1.1. *FFHB, 2E:* Page 228.

Question #44. The major natural factor affecting a fire stream is the wind and wind direction. Gravity and air friction are also natural factors, especially the farther the travel distance of a stream. NFPA 1001: 5.3 NFPA 6.3. *FFHB, 2E:* Page 295.

Question #45. Piercing nozzles were originally designed to penetrate the skin of aircraft and now have been modified to pierce through buildings walls and floors. NFPA 1001: 5.3 NFPA 6.3. *FFHB, 2E:* Page 286.

Question #46. Understanding of basic chemical and physical properties is important for the health and safety of emergency responders. NFPA 472: 5.2.2 - 5.2.4. *FFHB, 2E:* Page 800.

Question #47. A communications center may also be referred to as a PSAP, or public safety answering point. NFPA 1001: 5.2.1. *FFHB, 2E:* Page 48.

Question #48. First responders should have a basic awareness of the threat of terrorism and basic response actions. NFPA 472: 4.2.1. *FFHB, 2E:* Page 929.

Question #49. The rate-of-rise heat detector measures temperature increases above a predetermined rate. NFPA 1001: 6.5.1. *FFHB, 2E:* Page 310.

Question #50. Nozzle reaction is the force of nature that makes the nozzle move in the opposite direction of the water flow. NFPA 1001: 5.3 NFPA 6.3. *FFHB, 2E:* Page 282.

Question #51. Specific to cleaning and maintenance, the manufacturer will provide cleaning instructions and precautions, including a warning that the user should not wear equipment that is not thoroughly clean and dry. NFPA 1001: 5.1.1.2. *FFHB, 2E:* Page 136.

Question #52. Static pressure is the pressure in the system with no hydrants or water flowing. NFPA 1001: 5.3.15. *FFHB, 2E:* Page 214.

Question #53. Deep wells may penetrate through several layers of water before finding an aquifer. They are a more predictable water source with less chance of contamination. NFPA 1001: 5.3.15. *FFHB, 2E:* Page 204.

Question #54. These detectors use a radioactive element that emits ions into a chamber. NFPA 1001: 6.5.1. *FFHB, 2E:* Page 311.

Question #55. The portable tank is erected or inflated at the dump site in a location where the loaded tenders can offload their water and an engine can place its suction hose into the tank to draft and supply the water to attack engines. NFPA 1001: 5.3.15. *FFHB, 2E:* Page 210.

Question #56. Computer-aided dispatch systems are used in emergency communications centers to track the locations of emergency incidents and what units have been assigned to respond to emergency incidents. NFPA 1001: 5.2.1. *FFHB, 2E:* Page 50.

Question #57. The four elements of a fire stream are the pump, water, hose, and nozzle. NFPA 1001: 5.3 NFPA 6.3. *FFHB, 2E:* Page 281.

Question #58. Atmospheres with oxygen concentrations below 19.5 percent are classified as oxygen-deficient atmospheres. NFPA 1001: 5.3.1. *FFHB, 2E:* Page 145.

Question #59. Class D fires involve combustible metals and alloys such as magnesium, sodium, lithium, and potassium. NFPA 1001: 5.3.16. *FFHB, 2E:* Page 186.

Question #60. The Becket bend is utilized to tie ropes of equal diameter together. The double Becket bend is used most often when tying ropes of unequal diameter. NFPA 1001: 5.1.1.1. *FFHB, 2E:* Page 429.

Question #61. These cylinders have an unlimited service life as long as they pass a hydrostatic test every five years. NFPA 1001: 5.3.1. *FFHB, 2E:* Page 154.

Question #62. First responders should have a basic awareness of the threat of terrorism and basic response actions. NFPA 472: 4.2.1. *FFHB, 2E:* Page 920.

Question #63. Fire hose is made in three types of construction: wrapped, braided, and woven. NFPA 1001: 5.3. *FFHB, 2E:* Page 221.

Question #64. Understanding of basic chemical and physical properties is important for the health and safety of emergency responders. NFPA 472: 5.2.2 - 5.2.4. *FFHB, 2E:* Page 795.

Question #65. Drafting water from a lake or the sea is accomplished by taking advantage of the atmospheric pressure. Creating a partial vacuum or low atmospheric pressure area inside a pump causes the atmospheric pressure on the waters surface to force the water up the suction hose and into the pump, which adds pressure and pumps it out. NFPA 1001: 5.3.15. *FFHB, 2E:* Page 212.

Question #66. A hydrocarbon is an organic compound containing only carbon and hydrogen. NFPA 1001: 5.3; 6.3. *FFHB, 2E:* Page 79.

Question #67. Vapor pressure is the measurable amount of pressure being exerted against a confined container by a liquid substance as it converts to a gas. NFPA 1001: 5.3; 6.3. *FFHB, 2E:* Page 86.

Question #68. Responders should understand basic recognition and identification skills. NFPA 472: 5.2.2 - 5.2.4. *FFHB, 2E:* Page 757.

Question #69. Hydraulics is the study of fluids at rest and in motion, which describes the flow pattern of water supply and fire streams. NFPA 1001: 5.3 NFPA 6.3. *FFHB, 2E:* Page 290.

Question #70. Risk management is the process of minimizing the chance, degree, or probability of damage, loss, or injury. NFPA 1001: 6.1.1.1. *FFHB, 2E:* Page 108.

Question #71. The communication process includes four basic elements: receive, understand, record, and communicate. NFPA 1001: 5.2.1. *FFHB, 2E:* Page 47.

Question #72. Understanding of basic chemical and physical properties is important for the health and safety of emergency responders. NFPA 472: 5.2.2 - 5.2.4. *FFHB, 2E:* Page 795.

Question #73. Transfer of command should occur during a face-to-face meeting; but under extreme conditions, transfer may be accomplished by radio or telephone. NFPA 1001: 6.1.1.1. *FFHB, 2E:* Page 37.

Question #74. Wall hydrants are hydrants mounted on the wall of a building after the water line has been run into the building. NFPA 1001: 5.3.15. *FFHB, 2E:* Page 209.

Question #75. The combustion process produces toxic gases and irritants that affect both the short- and long-term health of firefighters operating in hazardous environments. NFPA 1001: 5.3.1. *FFHB, 2E:* Page 146.

Question #76. When using any nontrunked two-way radio, it is important to depress and hold the push-to-talk button at least two seconds before talking to avoid clipping the first part of the message. NFPA 1001: 5.2.3. *FFHB, 2E:* Page 65.

Question #77. Master stream devices are capable of flowing more than 350 gpm, and some have capabilities of many thousands of gallons per minute. This is the artillery of the fire service and is used when large volumes are required. NFPA 1001: 5.3 NFPA 6.3. *FFHB, 2E:* Page 289.

Question #78. This pressure (1 1/2 to 2 psi, depending on the manufacturer), which is slightly above atmospheric pressure, also helps maintain face piece seals. NFPA 1001: 5.3.1. *FFHB, 2E:* Page 144.

Question #79. A groove is notched into coupling lugs to help firefighters align the Higbee cuts.This notch is called the Higbee indicator. NFPA 1001: 5.1.1.1. *FFHB, 2E:* Page 226.

Question #80. The difference between this standard and the OSHA regulations is that a government authority (city, town, county, or state), called the authority having jurisdiction (AHJ) , must adopt the standard as policy for the fire department. NFPA 1001: 5.3.1. *FFHB, 2E:* Page 148.

Question #81. A secondary water source is a fire department Siamese connection, which allows pumpers to supplement the water supply. NFPA 1001: 6.5.1. *FFHB, 2E:* Page 322.

Question #82. PASS, the four steps for using a fire extinguisher: Pull the pin, aim the nozzle, squeeze the handle, and sweep the base of the fire. NFPA 1001: 5.3.16. *FFHB, 2E:* Page 195.

Question #83. The National Fire Protection Association requires that PASS devices integrated with SCBA activate automatically when the SCBA air supply is turned on. NFPA 1001: 5.1.1.2. *FFHB, 2E:* Page 136.

Question #84. A work uniform that meets NFPA standards is not designed to protect the wearer from IDLH atmospheres, but can add another layer of reasonable protection under wildland, proximity, and structural ensembles. NFPA 1001: 5.1.1.2. *FFHB, 2E:* Page 136.

Question #85. A TDD is a device that allows citizens to communicate with a telecommunicator through the use of a keyboard over telephone circuits instead of voice communications. NFPA 1001: 5.2.1; 5.2.2. *FFHB, 2E:* Page 58.

Question #86. Mobile water supply apparatus, according to NFPA 1901, Standard for Automotive Fire Apparatus, must have a minimum of a 1000-gallon water tank. NFPA 1001: 5.3.15. *FFHB, 2E:* Page 31.

Question #87. An oxidizer is a catalyst in the breakdown of molecules and possesses a chemical property that can pull apart a molecule and break apart the bond that previously existed. NFPA 1001: 5.3; 6.3. *FFHB, 2E:* Page 81.

Question #88. The National Institute of Occupational Safety and Health is mandated by legislation to investigate all incident-related firefighter fatalities. NFPA 1001: 6.1.1.1. *FFHB, 2E:* Page 110.

Question #89. A 2-A extinguisher will put out twice the fire of a 1-A. NFPA 1001: 5.3.16. *FFHB, 2E:* Page 184.

Question #90. Within an Incident Management System, there are four section chiefs: operations, logistics, planning, and finance/administration. NFPA 1001: 6.1.1.1. *FFHB, 2E:* Page 39-40.

Question #91. Understanding of basic chemical and physical properties is important for the health and safety of emergency responders. NFPA 472: 5.2.2 - 5.2.4. *FFHB, 2E:* Page 700.

Question #92. First responders should have a basic awareness of the threat of terrorism and basic response actions. NFPA 472: 4.2.1. *FFHB, 2E:* Page 928.

Question #93. Understanding of basic chemical and physical properties is important for the health and safety of emergency responders. NFPA 472: 5.2.2 - 5.2.4. *FFHB, 2E:* Page 797.

Question #94. Vapor density describes the weight of a gas as compared to normal air and is identified as a number. NFPA 1001: 5.3; 6.3. *FFHB, 2E:* Page 87.

Question #95. Company officers are supervisory-level positions and are responsible for both firefighters and administrative duties. NFPA 1001: 5.1.1.1. *FFHB, 2E:* Page 28.

Question #96. The Occupational Safety and Health Administration, which is a part of the Department of Labor, is responsible for the enforcement of safety-related regulations in the workplace. NFPA 1001: 6.1.1.1. *FFHB, 2E:* Page 109.

Question #97. An encoder is a device that converts the entered code into paging codes, which then activate a variety of paging devices. NFPA 1001: 5.2.1. *FFHB, 2E:* Page 60.

Question #98. The NFPA standard entitled Standard for Fire Fighter Professional Qualifications is also known as NFPA 1001. NFPA 1001: All. *FFHB, 2E:* Page 26.

Question #99. Occasionally water should be flowed through stored sections to prevent the liner from drying and cracking, and then it should be re-dried and stored. NFPA 1001: 5.5.4. *FFHB, 2E:* Page 224.

Question #100. The legal term standard of care means for every emergency medical incident, an emergency responder should treat the patient in the same manner as would another emergency responder with the same training. NFPA 1001: 4.3. *FFHB, 2E:* Page 694.

Phase One, Exam Two

1. Ladders can have special uses, such as for elevated hose streams and ventilation fan supports.

   a. True

   b. False

2. Loads can be divided into two broad categories as it relates to building construction: dead loads and live loads.

   a. True

   b. False

3. All overhead obstructions must be considered when placing a ladder.

   a. True

   b. False

4. A primary search is a slow, thorough search.

   a. True

   b. False

5. Energy conservation can lead to structural failure, flashover, and backdraft.

   a. True

   b. False

6. The call-taking process consists of receiving a report, intervening, and referral or dispatch composition.

   a. True

   b. False

7. Steel does not lose its strength as the temperatures in fire conditions increase.

   a. True

   b. False

8. The cutting torch is a great resource to firefighters as it affords quick results and does not require any specialized training.

   a. True

   b. False

9. Carbon monoxide combines with blood almost 100 times easier than oxygen.

   a. True

   b. False

10. Whenever two firefighters enter an involved structure, there should be two firefighters outside ready to enter and assist should a problem develop.

    a. True

    b. False

11. Command and control agreements are prearranged, written agreements of the type and amount of assistance one jurisdiction will provide to another in the event of a large-scale fire or disaster.

    a. True

    b. False

12. Being "macho" is a trait that has resulted in firefighter illness.

    a. True

    b. False

13. Simply stated, take any side of the fire tetrahedron away and the fire will go out.

    a. True

    b. False

14. The advantage of using synthetic rope in the fire service is that it does not have to be inspected.

    a. True

    b. False

15. A written standard operating procedure (SOP) is an example of a formal procedure.

    a. True

    b. False

16. A chemical with a vapor pressure of less than 1 is heavier than air and will tend to collect in low-lying areas such as depressions, basements, cellars, sewers, ravines, or the like.

    a. True

    b. False

17. Rope can be washed in any washing machine as long as it is placed in a canvas bag.

    a. True

    b. False

18. Tempered glass is also known as safety glass.

    a. True

    b. False

19. The major concern of a vehicle fire is combating Class A fires that erupt in the engine compartment.

    a. True

    b. False

20. To stabilize the head and neck of a patient, a firefighter places one hand on either side of the patient's head and holds firmly so that the head and neck are in a straight line with the body. The purpose of stabilizing the head and neck is to keep them from moving so that if there is a neck injury, no further damage will be done to the neck or spinal cord.

    a. True
    b. False

21. The STEL is an average of _____ for a chemical exposure.

    a. 15 minutes
    b. 8 hours
    c. 24 hours
    d. 1 hour

22. When a firefighter is performing a task that was not assigned to him or performing a task alone, he is said to be _____.

    a. accountable
    b. freelancing
    c. performing as trained
    d. fit for duty

23. The state of conversion from a liquid to a gas under normal atmospheric conditions is called _____.

    a. vapor density
    b. evaporation
    c. equilibrium
    d. diffusion

24. Which is not a type of water extinguisher?

    a. Pump type
    b. Pressurized
    c. Pressurized loaded stream
    d. Cartridge-operated type

25. A ladder belt is also known as a _____ harness.

    a. Class 1
    b. Class 2
    c. Class L
    d. Class A

26. _____ are used to contain the water dripping through a ceiling until a system is set up to remove the water from the building.

    a. Squeegees
    b. Catch-alls
    c. Water chutes
    d. Floor runners

27. _____ operations consist of breaching walls, floors, ceilings, and dead air spaces in order to confirm that fire is not present in these areas.
    a. Fire investigation
    b. Overhaul
    c. Salvage
    d. Point-of-origin

28. A(n) _____ plan is a predetermined response plan of apparatus and personnel for specific types of incidents and specific locations.
    a. apparatus
    b. alarm
    c. deployment
    d. database

29. The most fatalities occur in _____ fires.
    a. commercial
    b. residential
    c. industrial
    d. aircraft/rescue

30. What type of door can be used as a required exit in a building?
    a. Swinging door
    b. Revolving door
    c. Overhead door
    d. Sliding door

31. A combination attack is _____.
    a. a blend of direct and indirect attack
    b. not a recognized procedure under NFPA
    c. strictly a defensive attack
    d. strictly an offensive attack

32. _____ is that point where the need for outside heat application ceases and the ability for the material to sustain combustion comes from the heat generation of the material itself.
    a. Ignition
    b. Flammability
    c. Rollover
    d. Decay

33. A firefighter discovers that a facility with outdoor storage of hazardous materials has a six-inch curb around the storage area. What is this called?
    a. Primary containment
    b. Primary hazard control
    c. Secondary containment
    d. Secondary hazard control

34. What is not a route of exposure?

    a. Inhalation

    b. Substernal

    c. Absorption

    d. Ingestion

35. The _____ has a movable shackle that locks into the body of the lock and is used to secure a door or gate using a hasp or chain.

    a. rim lock

    b. padlock

    c. tubular lock

    d. mortise lock

36. Which is not a side on the wildland fire triangle?

    a. Topography

    b. Fuel

    c. Heat

    d. Weather

37. The DOT ERG lists a small spill that is less than _____.

    a. a tank truck

    b. a small rail car

    c. 1,000-gallon tank

    d. 55-gallon drum

38. Flashover survival skill can be defined as _____.

    a. recognition and an indirect interior attack

    b. recognition and waiting for clues

    c. recognition and avoidance

    d. recognition and aggressive attack

39. A patient entrapped by sand in a below grade situation whose head is not buried is considered _____.

    a. safe

    b. in danger

    c. stable

    d. none of the above

40. _____ can be defined as a continuous mental evaluation of firefighters' immediate environments, facts, and probabilities.

    a. Freelancing

    b. Personal assignment

    c. Personal accountability

    d. Personal size-up

41. A _____ is a communications device that has no information processing capabilities.

   a. mobile data terminal

   b. mobile radio

   c. mobile support vehicle

   d. mobile data computer

42. _____ is a term describing actions taken to eliminate a hazard or make a hazard less severe or less likely to cause harm. It is considered a proactive action.

   a. Mitigation

   b. Preparedness

   c. Intervention

   d. Risk management

43. Firefighters are performing an initial assessment on an unconscious adult patient. Where will they locate the pulse to assess this patient's circulation?

   a. The carotid pulse

   b. The radial pulse

   c. The brachial pulse

   d. The femoral pulse

44. What incident command designation is established to maintain span of control over a number of divisions, sectors, and/or groups?

   a. A strike team

   b. A branch

   c. A section

   d. A task force

45. A(n) _____ ladder is a collapsible straight ladder typically used to access tight or narrow spaces such as attics.

   a. folding

   b. A-frame combination

   c. hook

   d. extension

46. _____ are very helpful salvage tools when the need exists to move large amounts of water on a flat surface.

   a. Submersible pumps

   b. Squeegees

   c. Salvage covers

   d. Floor runners

47. Heavy-duty padlocks can be opened with a _____.
    a. chainsaw
    b. Halligan
    c. duck-bill lock breaker
    d. bird-leg cramp tool

48. What stage of fire is reached when all the fuel has been consumed?
    a. Ignition
    b. Growth
    c. Fully developed
    d. Decay

49. When a firefighter is involved in a water rescue, the last method of rescue is
    _____.
    a. reach
    b. throw
    c. row
    d. go

50. A fitness program that includes _____ helps firefighters build muscle in order
    to make the demands of firefighting easier on the body.
    a. cardiovascular conditioning
    b. flexibility improvement
    c. resistance training
    d. core balancing

51. A dam that allows for clean water to flow over the dam and collect the chemical at the
    base of the dam is called a(n) _____.
    a. underflow dam
    b. dike
    c. absorbent
    d. overflow dam

52. A rescue knot _____.
    a. is a standardized knot used for rescue in the fire service
    b. is a knot used on oneself, or a conscious or unconscious patient
    c. should only be tied on an emergency scene
    d. is no longer acceptable

53. The potential hazards section in the DOT ERG response guides lists the
    _____.
    a. most dangerous hazard first
    b. most dangerous hazard last
    c. highlights the hazard
    d. is in random order

54. During fire safety inspections, firefighters should check fire sprinkler systems to ensure that all water supply vales are _____ and secured.

   a. closed

   b. hidden

   c. open

   d. capped

55. Wooden tool handles should be _____.

   a. cleaned with soap and water

   b. painted

   c. varnished

   d. painted and then varnished

56. What is a process for managing the short- and long-term effects of critical incident stress reactions?

   a. Critical Incident Stress Management

   b. Work Hardening

   c. Critical Incident Stress Debriefing

   d. Employee Assistance Program

57. A(n) _____ load is a load that is in motion when applied.

   a. impact

   b. concentrated

   c. undesigned

   d. distributed

58. Which type of exposure causes immediate effects?

   a. Acute

   b. Chronic

   c. Afflicted

   d. Carcinogenic

59. What kind of exposure presents delayed effects?

   a. Acute

   b. Chronic

   c. Arbitrary

   d. Afflicted

60. Rescues involving escalators usually involve the _____.

   a. steps

   b. landing plate

   c. glass sides

   d. none of the above

61. The first "E" in the acronym RECEO is for _____.

   a. extinguishment

   b. exhume

   c. exposures

   d. extrication

62. What is a defined program that offers professional mental health and other health services to employees?

   a. Critical Incident Stress Management

   b. Work Hardening

   c. Critical Incident Stress Debriefing

   d. Employee Assistance Program

63. A Class A fire involves _____.

   a. ordinary combustibles

   b. flammable and combustible liquids, gases, and greases

   c. combustible metals and alloys

   d. energized electrical equipment

64. When a concentration of a gas falls into the range where it can ignite, it is said to be within its _____ limits.

   a. flammable

   b. explosive

   c. a and b

   d. none of the above

65. The legal term _____ means for every emergency medical incident, an emergency responder should treat the patient in the same manner as would another emergency responder with the same training.

   a. consent

   b. abandonment

   c. standard of care

   d. implied care

66. A _____ radio system uses two frequencies per channel, transmitting outgoing messages on one and receiving incoming messages on the other.

   a. simplex

   b. multisite

   c. secondary

   d. duplex

67. At best, the survival time of a firefighter in full PPE and SCBA involved in a flashover is _____.
    a. 5-10 seconds
    b. 10-15 seconds
    c. 15-20 seconds
    d. none (A firefighter cannot survive a flashover.)

68. Short-term for public protection options is regarded as _____.
    a. years
    b. minutes to hours
    c. days
    d. weeks

69. A BLEVE is a boiling liquid _____ vapor explosion.
    a. exothermic
    b. evaporating
    c. extracting
    d. expanding

70. A _____ is used to secure the firefighter to the ladder when both hands must be used to perform a task and a ladder belt is not available.
    a. rung lock
    b. halyard
    c. leg lock
    d. carry lock

71. Who has overall responsibility for all actions at a hazardous materials incident?
    a. Incident commander
    b. Police
    c. HazMat officer
    d. Emergency management department

72. PPV evacuation of smoke is successful for distances of up to _____ feet.
    a. 1,000
    b. 500
    c. 250
    d. 100

73. Confined spaces are found mainly _____.
    a. above grade
    b. below grade
    c. at grade
    d. all of the above

74. To force entry into a folding overhead garage door, you should use which of the following?

    a. Access the securing device and unlock it.

    b. Gain access through the lock entry.

    c. Break a panel and pull the release cord.

    d. All of the above are correct.

75. Which of the following is not one of the modes of heat transfer?

    a. Conduction

    b. Mechanical

    c. Convection

    d. Radiation

76. OSHA requires two important positions be assigned at a hazardous materials incident. One is the incident commander. What is the other position?

    a. Remediation officer

    b. Entry officer

    c. Degradation officer

    d. Safety officer

77. In one common type of accountability system, firefighters give a name chip to the company officer or team leader, who places all of the team names on a card or _____.

    a. tag

    b. assignment board

    c. passport

    d. clipboard

78. How many firefighters are needed to perform a salvage cover roll?

    a. One

    b. Two

    c. Three

    d. Four

79. Which is not included in the DOT ERG?

    a. Placard information

    b. Response guides

    c. Advanced response information

    d. Isolation distances

80. What government agency is closely involved in the allocation and monitoring of radio frequencies to public safety agencies?

    a. The National Emergency Number Association (NENA)

    b. The Federal Communications Commission (FCC)

    c. The Federal Emergency Management Agency (FEMA)

    d. The National Fire Protection Association (NFPA)

81. The permissible exposure limit is an average of _____ for a chemical exposure.

    a. 15 minutes

    b. 8 hours

    c. 24 hours

    d. 1 hour

82. The use of the bowline knot has been greatly reduced because of _____.

    a. failure

    b. NFPA standardization of rope fibers

    c. synthetic rope

    d. natural fiber rope

83. Which of the following salvage tools would be helpful in a situation involving a flooded basement?

    a. Salvage covers

    b. Catch-alls

    c. Water vacuums

    d. Submersible pumps

84. The most important single forcible entry tool used in the fire service is the _____.

    a. axe

    b. Halligan tool

    c. Pulaski tool

    d. maul

85. A(n) _____ load is a load that is applied to a small area.

    a. impact

    b. concentrated

    c. design

    d. distributed

86. Although the use of hot, warm, and cold zones is recommended, what zone or area should be immediately established by the first-arriving responders?

    a. Cold sector

    b. Isolation zone

    c. Rehab sector

    d. Warm zone

87. A _____ is any combination of single resources assembled for an assignment.

    a. task force

    b. crew

    c. group

    d. division

88. The preferred method of cleaning a rope is _____.
    a. a washing machine
    b. a rope washer
    c. hand washing
    d. recommended by the manufacturer

89. An important component of a successful incident management system is the use of
    _____ terminology.
    a. unique
    b. complicated
    c. common
    d. command

90. The distance between the ground and the ladder's point of contact with a structure is
    called the _____ length.
    a. extended
    b. working
    c. raising
    d. carry

91. Which is not a part or portion of a knot?
    a. The working end
    b. The standing part
    c. The standing end
    d. The running end

92. What is a formal gathering of incident responders to help address the stress from
    a given incident?
    a. Critical Incident Stress Management
    b. Vicarious Experience
    c. Critical Incident Stress Debriefing
    d. Employee Assistance Program

93. For what is the orange section of the DOT ERG?
    a. Alphabetical listing of chemicals
    b. Listing of placards
    c. List of UN numbers
    d. Response guides

94. When ventilating a sealed, climate-controlled building, a firefighter should consider
    _____.
    a. HVAC
    b. PPV
    c. NPV
    d. hydraulic ventilation

95. Signs of a potential backdraft include _____.

    a. smoke-stained windows

    b. smoke under pressure

    c. extreme heat

    d. all of the above

96. A _____ is a structural element that delivers loads perpendicular to its length.

    a. beam

    b. column

    c. connection

    d. wall

97. For real events, after pipe bombs, what is the next leading category of threat agent?

    a. Ricin

    b. Anthrax

    c. RDD

    d. Nerve agents

98. The first record of a truly organized fire department began with actions taken by the _____ to protect their capital.

    a. Americans

    b. British

    c. Romans

    d. Germans

99. A _____ load is a load that is applied offset to an element, causing a twisting stress to the material.

    a. impact

    b. eccentric

    c. axial

    d. torsion

100. What type of door can be used as a required exit in a building?

    a. Swinging door

    b. Revolving door

    c. Overhead door

    d. Sliding door

Phase I, Exam II: Answers to Questions

1. T	26. B	51. D	76. D
2. T	27. B	52. A	77. C
3. T	28. C	53. A	78. B
4. F	29. B	54. C	79. C
5. T	30. A	55. A	80. B
6. F	31. A	56. A	81. B
7. F	32. A	57. A	82. C
8. F	33. C	58. A	83. D
9. F	34. B	59. B	84. B
10. T	35. B	60. B	85. B
11. F	36. C	61. C	86. B
12. T	37. D	62. D	87. A
13. T	38. C	63. A	88. D
14. F	39. B	64. C	89. C
15. T	40. D	65. C	90. B
16. F	41. A	66. D	91. C
17. F	42. A	67. B	92. C
18. F	43. A	68. B	93. D
19. F	44. B	69. D	94. A
20. T	45. A	70. C	95. D
21. A	46. B	71. A	96. A
22. B	47. C	72. A	97. A
23. B	48. D	73. D	98. C
24. D	49. D	74. D	99. D
25. A	50. C	75. B	100. A

Phase I, Exam II:
Rationale & References for Questions

Question #1. Ladders can have special uses, such as for elevated hose streams and ventilation fan supports. NFPA 1001: 5.36; 5.3.9; 5.3.10; 5.3.11; 5.3.12. *FFHB, 2E:* Page 388; 390.

Question #2. Loads can be divided into two broad categories as it relates to building construction: dead loads and live loads. NFPA 1001: 5.3.10; 6.3.2. *FFHB, 2E:* Page 342.

Question #3. All overhead obstructions must be considered when placing a ladder. NFPA 1001: 5.36; 5.3.9; 5.3.10; 5.3.11; 5.3.12. *FFHB, 2E:* Page 390.

Question #4. They search the areas that are most likely to have victims in a rapid, but thorough, manner. NFPA 1001: 5.3.9. *FFHB, 2E:* Page 465.

Question #5. These tight construction practices lead to hotter fires, early failure of structural components, and greater incidences of flashover and backdraft. NFPA 1001: 5.3.11. *FFHB, 2E:* Page 550.

Question #6. The call-taking process consistes of receiving a report, interviewing, and referral or dispatch composition. NFPA 1001: 5.2.1; 5.2.2. *FFHB, 2E:* Page 51.

Question #7. Steel loses strength as the temperatures in fire conditions increase. NFPA 1001: 5.3.10; 6.3.2. *FFHB, 2E:* Page 349.

Question #8. Use of cutting torches requires specialized training in addition to following manufacturer's recommendations and department procedures. NFPA 1001: 5.3.4; 6.3.2. *FFHB, 2E:* Page 517.

Question #9. In fact, CO combines with blood almost 218 times easier than oxygen. NFPA 1001: 5.3.1. *FFHB, 2E:* Page 147.

Question #10. In addition to this search team, a minimum of two firefighters must be standing by immediately outside in full protective clothing and SCBA with a charged hoseline (although the charged hoseline is not required, it is a good logical practice) ready to come in and assist the search team should a problem develop. NFPA 1001: 5.3.9. *FFHB, 2E:* Page 462.

Question #11. Mutual aid or assistance agreements are prearranged, written agreements of the type and amount of assistance one jurisdiction will provide to another in the event of a large-scale fire or disaster. NFPA 1001: 5.1.1.1. *FFHB, 2E:* Page 36.

Question #12. Injuries and illnesses have been suffered by the firefighter who fails to properly don and secure PPE—usually because the wearer was trying to create a "macho" image or skipped complete donning in the haste to perform a task. NFPA 1001: 5.1.1.2. *FFHB, 2E:* Page 125.

Question #13. These elements are commonly suppressed by removing one side of the fire tetrahedron. Page 596. *FFHB, 2E:* Page 125.

Question #14. As with any emergency service tool, all ropes must be inspected and properly maintained to ensure they are in good shape for use during an emergency incident. NFPA 1001: 5.1.1.1. *FFHB, 2E:* Page 439.

Question #15. A written standard operating procedure (SOP) is an example of a formal procedure. NFPA 1001: 6.1.1.1. *FFHB, 2E:* Page 111.

Question #16. A chemical with a vapor pressure of greater than 1 is heavier than air and will tend to collect in low-lying areas such as depressions, basements, cellars, sewers, ravines, or the like. NFPA 1001: 5.3; 6.3. *FFHB, 2E:* Page 87.

Question #17. Only a front-loading machine (with a glass window) should be used. NFPA 1001: 5.5.3. *FFHB, 2E:* Page 442.

Question #18. Laminated glass is also known as safety glass. NFPA 1001: 5.3.4; 6.3.2; 6.4.1. *FFHB, 2E:* Page 542.

Question #19. These are Class B fires as opposed to Class A fires. NFPA 1001: 5.3.7. *FFHB, 2E:* Page 600.

Question #20. To stabilize the head and neck of a patient, a firefighter places one hand on either side of the patient's head and holds firmly so that the head and neck are in a straight line with the body. The purpose of stabilizing the head and neck is to keep them from moving so that if there is a neck injury, no further damage will be done to the neck or spinal cord. NFPA 1001: 4.3. *FFHB, 2E:* Page 762.

Question #21. Understanding toxicology and health effects is important. NFPA 472: 5.2.2; 5.2.3. *FFHB, 2E:* Page 838.

Question #22. When a firefighter is performing a task that was not assigned to him or performing a task alone, he is said to be freelancing. NFPA 1001:5.3; 6.3. *FFHB, 2E:* Page 725.

Question #23. The state of conversion from a liquid to a gas under normal atmospheric conditions is called evaporation. NFPA 1001: 5.3; 6.3. *FFHB, 2E:* Page 89.

Question #24. 1.Water type for Class A fires: a.Pump type b.Pressurized water c.Pressurized loaded stream NFPA 1001: 5.3.16. *FFHB, 2E:* Page 188.

Question #25. This harness is commonly known as a ladder belt, and the hook is used to secure a firefighter to a ladder. NFPA 1001: 6.4.2. *FFHB, 2E:* Page 491.

Question #26. Catch-alls are used to contain the water dripping through a ceiling until a system is set up to remove the water from the building. NFPA 1001: 5.3.14. *FFHB, 2E:* Page 645.

Question #27. Overhaul operations consist of breaching walls, floors, ceilings, and dead air spaces in order to confirm that fire is not present in these areas. NFPA 1001: 5.3.13. *FFHB, 2E:* Page 649.

Question #28. A deployment plan is a predetermined response plan of apparatus and personnel for specific types of incidents and specific locations. NFPA 1001: 5.2.3. *FFHB, 2E:* Page 59.

Question #29. More people die in residential fires than in any other type of fire in the United States. NFPA 1001: 5.1.1.1. *FFHB, 2E:* Page 611.

Question #30. Only a swinging door can be used as a required exit in a building. NFPA 1001: 5.5.1. *FFHB, 2E:* Page 664.

Question #31. This attack is a blend of both the direct and indirect methods. NFPA 1001: 5.3.10. *FFHB, 2E:* Page 604.

Question #32. Ignition is that point where the need for outside heat application ceases and the ability for the material to sustain combustion comes from the heat generation of the material itself. NFPA 1001: 5.3; 6.3. *FFHB, 2E:* Page 91.

Question #33. A firefighter discovers that a facility with outdoor storage of hazardous materials has a six-inch curb around the storage area. This is called secondary containment and is meant to contain the stored liquid if it should escape from its container. NFPA 1001: 5.5.1. *FFHB, 2E:* Page 670.

Question #34. Understanding toxicology and health effects is important. NFPA 472: 5.2.2; 5.2.3. *FFHB, 2E:* Page 835.

Question #35. This type of locking device has a movable shackle that locks into the body of the lock and is used to secure a door or gate using a hasp or chain. NFPA 1001: 5.3.4; 6.3.2. *FFHB, 2E:* Page 529.

Question #36. Figure 19-5 NFPA 1001: 5.3.19. *FFHB, 2E:* Page 596.

Question #37. Specific chemical information is important for a safe and informed response. NFPA 472: 4.2.1 - 4.2.2. *FFHB, 2E:* Page 817.

Question #38. The best survival skill here is recognition and avoidance. NFPA 1001: 5.3.11. *FFHB, 2E:* Page 556.

Question #39. This is why a victim succumbs to asphyxiation even though the head is not buried. NFPA 1001: 6.4.2. *FFHB, 2E:* Page 497.

Question #40. Personal size-up can be defined as a continuous mental evaluation of firefighters' immediate environments, facts, and probabilities. NFPA 1001:5.3; 6.3. *FFHB, 2E:* Page 725.

Question #41. A mobile data terminal is a communications device that has no information processing capabilities. NFPA 1001: 5.2.3. *FFHB, 2E:* Page 61-62.

Question #42. Mitigation is a term describing actions taken to eliminate a hazard or make a hazard less severe or less likely to cause harm. It is considered a proactive action. NFPA 1001: 6.1.1.1. *FFHB, 2E:* Page 110.

Question #43. NFPA 1001: 4.3. *FFHB, 2E:* Page 763.

Question #44. As an incident expands, branches are established to maintain span of control over a number of divisions, sectors, and/or groups. NFPA 1001: 6.1.1.1. *FFHB, 2E:* Page 42.

Question #45. A folding ladder is a collapsible straight ladder typically used to access tight or narrow spaces such as attics. NFPA 1001: 5.36; 5.3.9; 5.3.10; 5.3.11; 5.3.12. *FFHB, 2E:* Page 379.

Question #46. Squeegees are very helpful salvage tools when the need exists to move large amounts of water on a flat surface. NFPA 1001: 5.3.14. *FFHB, 2E:* Page 635.

Question #47. The duck-bill lock breaker is designed to break open heavy-duty padlocks. NFPA 1001: 5.3.4. *FFHB, 2E:* Page 518.

Question #48. When the point at which all fuel has been consumed is reached, the fire will begin to diminish in size. This is the decay stage. NFPA 1001: 5.3; 6.3. *FFHB, 2E:* Page 92.

Question #49. If none of the preceding options is possible, the absolute last method is for the rescuer to enter the water. NFPA 1001: 6.4.2. *FFHB, 2E:* Page 493.

Question #50. A fitness program that includes resistance training helps firefighters build muscle in order to make the demands of firefighting easier on the body. NFPA 1001:5.3; 6.3. *FFHB, 2E:* Page 727.

Question #51. Responders should understand the protective actions that are available to utilize. NFPA 472: 5.3.2 - 5.3.4. *FFHB, 2E:* Page 894.

Question #52. This rescue knot can be used on oneself, a conscious patient, or an unconscious patient.However, it is also important to practice the rescue knot on other people to simulate victims needing rescue. NFPA 1001: 5.1.1.1. *FFHB, 2E:* Page 437.

Question #53. Specific chemical information is important for a safe and informed response. NFPA 472: 4.2.1 - 4.2.2. *FFHB, 2E:* Page 812.

Question #54. During fire safety inspections, firefighters should check fire sprinkler systems to ensure that all water supply vales are open and secured. NFPA 1001: 5.5.1. *FFHB, 2E:* Page 666.

Question #55. During the pure combustion process, energy is released from an exothermic reaction as heat and light. NFPA 1001: 5.3.4. *FFHB, 2E:* Page 521.

Question #56. Critical Incident Stress Management is a process for managing the short- and long-term effects of critical incident stress reactions. NFPA 1001: 6.1.1.1. *FFHB, 2E:* Page 116.

Question #57. An impact load is a load that is in motion when applied. NFPA 1001: 5.3.10; 6.3.2. *FFHB, 2E:* Page 342-344.

Question #58. Understanding toxicology and health effects is important. NFPA 472: 5.2.2 ; 5.2.3. *FFHB, 2E:* Page 834.

Question #59. Understanding toxicology and health effects is important. NFPA 472: 5.2.2; 5.2.3. *FFHB, 2E:* Page 834.

Question #60. They usually involve a passenger getting feet caught in the area where the steps disappear into the landing plate or getting fingers caught under the moving hand rail. NFPA 1001: 6.4.2. *FFHB, 2E:* Page 504.

Question #61. The priority was first set in place a great many years ago by Lloyd Layman, with the following acronym: RECEO - Rescue, Exposures, Confinement, Extinguishment, Overhaul. NFPA 1001: 5.3.3. *FFHB, 2E:* Page 607.

Question #62. An Employee Assistance Program is a defined program that offers professional mental health and other health services to employees. NFPA 1001: 6.1.1.1. *FFHB, 2E:* Page 116.

Question #63. Class A fires involve ordinary combustibles such as wood, paper, cloth, plastics, and rubber. NFPA 5.3.16. *FFHB, 2E:* Page 185.

Question #64. When a concentration of a gas falls into the range where it can ignite, it is said to be within its flammable or explosive limits. NFPA 1001: 5.3; 6.3. *FFHB, 2E:* Page 90.

Question #65. The legal term standard of care means for every emergency medical incident, an emergency responder should treat the patient in the same manner as would another emergency responder with the same training. NFPA 1001: 4.3. *FFHB, 2E:* Page 694.

Question #66. A duplex radio system uses two frequencies per channel, transmitting outgoing messages on one and receiving incoming messages on the other. NFPA 1001: 5.2.3. *FFHB, 2E:* Page 64.

Question #67. At best, the survival time of a firefighter in bunker gear and breathing apparatus, fully encapsulated with gloves, hood, and helmet flaps down, is estimated to be between ten and fifteen seconds. NFPA 1001: 5.3.10. *FFHB, 2E:* Page 556.

Question #68. Understanding various protective actions is important for responder health and safety. NFPA 472:5.4.1. *FFHB, 2E:* Page 861.

Question #69. A BLEVE is a boiling liquid expanding vapor explosion. NFPA 1001: 5.3; 6.3. *FFHB, 2E:* Page 88.

Question #70. A leg lock is used to secure the firefighter to the ladder when both hands must be used to perform a task and a ladder belt is not available. NFPA 1001: 5.36; 5.3.9; 5.3.10; 5.3.11; 5.3.12. *FFHB, 2E:* Page 408.

Question #71. Understanding various protective actions is important for responder health and safety. NFPA 472: 5.4.2. *FFHB, 2E:* Page 859.

Question #72. Positive-pressure evacuation of smoke is successful for distances of up to 1,000 feet. NFPA 1001: 5.3.11. *FFHB, 2E:* Page 575.

Question #73. NFPA 1001: 6.4.2. *FFHB, 2E:* Page 498.

Question #74. A folding overhead garage door may be forced by any of several methods: Break a panel or window, reach in, and unlock the securing device. Pull the lock cylinder and utilize through-the-lock forcible entry. Automatic openers hold the door in the closed position. To disconnect the opener, break out a panel near the attachment mechanism, reach in with a tool to grab the release cord, and pull as shown. NFPA 1001: 5.3.4. *FFHB, 2E:* Page 525.

Question #75. The three modes of heat transfer are conduction, convection, and radiation. NFPA 1001: 5.3; 6.3. *FFHB, 2E:* Page 94.

Question #76. Understanding various protective actions is important for responder health and safety. NFPA 472: 5.4.2. *FFHB, 2E:* Page 859.

Question #77. In one common type of accountability system, firefighters give a name chip to the company officer or team leader, who places all of the team names on a card or passport. NFPA 1001:5.3; 6.3. *FFHB, 2E:* Page 725.

Question #78. Two firefighters are needed to roll a salvage cover. NFPA 1001: 5.3.14. *FFHB, 2E:* Page 637.

Question #79. Specific chemical information is important for a safe and informed response. NFPA 472: 4.2.1 - 4.2.2. *FFHB, 2E:* Page 807.

Question #80. The Federal Communications Commission (FCC) is closely involved in the allocation and monitoring of radio frequencies to public safety agencies. NFPA 1001: 5.2.3. *FFHB, 2E:* Page 63.

Question #81. Understanding toxicology and health effects is important. NFPA 472: 5.2.2; 5.2.3. *FFHB, 2E:* Page 838.

Question #82. Although it was the mainstay of the fire service knots for years, the advent of synthetic fiber ropes has greatly reduced the utilization of the bowline knot. NFPA 1001: 5.1.1.1. *FFHB, 2E:* Page 431.

Question #83. Submersible pumps are very good tools for flooded basements. NFPA 1001: 5.3.14. *FFHB, 2E:* Page 635.

Question #84. The original Halligan tool, designed by Hugh Halligan of the Fire Department of the City of New York, has proven to be the most important single forcible entry tool used in the fire service. NFPA 1001: 5.3.4. *FFHB, 2E:* Page 513.

Question #85. A concentrated load is a load that is applied to a small area. NFPA 1001: 5.3.10; 6.3.2. *FFHB, 2E:* Page 342-344.

Question #86. Understanding various protective actions is important for responder health and safety. NFPA 472:5.4.1. *FFHB, 2E:* Page 860.

Question #87. A task force is any combination of single resources assembled for an assignment. NFPA 1001: 6.1.1.1. *FFHB, 2E:* Page 42.

Question #88. As mentioned in the previous section, the best policy for cleaning is to follow the manufacturer's instructions. NFPA 1001: 5.1.1.1. *FFHB, 2E:* Page 441.

Question #89. An important component of a successful incident management system is the use of common terminology. NFPA 1001: 6.1.1.1. *FFHB, 2E:* Page 37.

Question #90. The distance between the ground and the ladder's point of contact with a structure is called the working length. NFPA 1001: 5.36; 5.3.9; 5.3.10; 5.3.11; 5.3.12. *FFHB, 2E:* Page 399.

Question #91. Note:It helps to think of the rope as having two ends and a part or portion between the ends. 1.The working end is the end of the rope utilized to tie the knot. 2.The standing part is between the working end and the running end. 3.The running end is used for work such as hoisting a tool. NFPA 1001: 5.1.1.1. *FFHB, 2E:* Page 425.

Question #92. Critical Incident Stress Debriefing is a formal gathering of incident responders to help address the stress from a given incident. NFPA 1001: 6.1.1.1. *FFHB, 2E:* Page 116.

Question #93. Specific chemical information is important for a safe and informed response. NFPA 472: 4.2.1 - 4.2.2. *FFHB, 2E:* Page 812.

Question #94. Heating, ventilation, and air-conditioning (HVAC) systems can be used effectively for ventilation in sealed, climate-controlled buildings. NFPA 1001: 5.3.11. *FFHB, 2E:* Page 563.

Question #95. Signs of a Potential Backdraft: Smoke-stained windows. Puffing of smoke at seams and cracks of windows and doors. Smoke pushing out under pressure. No visible flame or very dull red flame in the depth of the smoke. Heavy black smoke. Tightly sealed building. Large, open area structure (supermarket, bowling alley, department store). Can also be a large, open void (cockloft between-space of hanging ceiling). Extreme heat. NFPA 1001: 5.1.1.1. *FFHB, 2E:* Page 34.

Question #96. A beam is a structural element that delivers loads perpendicular to its length. NFPA 1001: 5.3.10; 6.3.2. *FFHB, 2E:* Page 346.

Question #97. First responders should have a basic awareness of the threat of terrorism and basic response actions. NFPA 472: 4.2.1. *FFHB, 2E:* Page 930.

Question #98. The first record of a truly organized fire department began with actions taken by the Romans to protect their capital. NFPA 1001: N/A. *FFHB, 2E:* Page 6.

Question #99. A torsion load is a load that is applied offset to an element, causing a twisting stress to the material. NFPA 1001: 5.3.10; 6.3.2. *FFHB, 2E:* Page 345.

Question #100. Only a swinging door can be used as a required exit in a building. NFPA 1001: 5.5.1. *FFHB, 2E:* Page 664.

Phase One, Exam Three

1. Complete and accurate communications center records should be maintained only on incidents involving medical calls or structure fires.

    a. True

    b. False

2. IDLH is the vernacular for Immediately Dangerous to Life due to Hazmat.

    a. True

    b. False

3. When climbing a ladder, firefighters should keep their eyes focused level and ahead, occasionally looking up the climbing path for hazards and to view the objective.

    a. True

    b. False

4. SCBA adds from 23 to 35 lbs to a firefighter's weight.

    a. True

    b. False

5. Hoselines can be hoisted using ropes, but they must be uncharged.

    a. True

    b. False

6. During fire operations, the potential for structural failure during a fire always exists.

    a. True

    b. False

7. PPE provides a minimum level of protection and should be considered the last resort of protection for firefighters and emergency responders.

    a. True

    b. False

8. For fire department operations, there are three standard methods of forcible entry:convectional, through-the-lock, and power tools.

    a. True

    b. False

9. Firefighters should avoid touching any radio antenna during transmission to avoid burns that can result from radio-frequency energy.

    a. True

    b. False

10. Inspection and maintenance is not a critical element that determines whether or not equipment is safe to operate.

   a. True

   b. False

11. K-tool use exposes the lock mechanism, which is then operated with the proper key tool.

   a. True

   b. False

12. The "total flooding system" is a fire protection system.

   a. True

   b. False

13. The wrong extinguisher can be as effective as the right extinguisher.

   a. True

   b. False

14. According to NFPA 1901, Standard for Automotive Fire Apparatus, aerial fire apparatus must have a minimum of 115 feet of ground ladders, including at least one attic ladder, two straightladders with folding roof hooks, and two extension ladders.

   a. True

   b. False

15. Once electrical power is disconnected from the airbag, it is no longer a threat to firefighters.

   a. True

   b. False

16. A bowline knot is the only knot that is recommended for tying webbing.

   a. True

   b. False

17. Fire flow capacity is the amount of projected water that a fire can contain.

   a. True

   b. False

18. Knockout speed is how fast the foam spreads across the surface of a fuel.

   a. True

   b. False

19. Hoseline advancement up a ladder should never be done with the line charged.

   a. True

   b. False

20. A lean-to collapse normally does not provide any chance of survival for victims.

   a. True

   b. False

21. What level of protective clothing can be used for splashes and help keep the respiratory hazards minimal?
    a. A
    b. B
    c. C
    d. D

22. Cross-use of rope for utility and life safety purposes is _____.
    a. recommended by NFPA
    b. cost-effective
    c. dangerous
    d. acceptable in some jurisdictions

23. Which of the following term describes something a fire investigator would look for to find a fire's point of origin?
    a. V pattern
    b. X pattern
    c. T pattern
    d. Y pattern

24. What is considered to be oxygen deficient?
    a. 20.9%
    b. 21%
    c. 19.5%
    d. 16%

25. The benefits of self-discipline applied to PPE completeness pay a dividend in the form of _____.
    a. a better self-image
    b. acclimation
    c. minor burns instead of major burns
    d. performance evaluation

26. The National Fire Protection Association (NFPA) created a _____ series of standards that specifically focuses on safety issues.
    a. 1200
    b. 1300
    c. 1500
    d. 1700

27. For a Class B extinguisher, _____.
    a. the number rating compares to the equivalent size of a Class A extinguisher
    b. the number rating compares approximately to the square footage to be extinguished
    c. the number rating has no bearing on the size of the fire, but the weight of the extinguisher
    d. the rating assumes an experienced operator

28. A _____ is a beam that supports other beams.
    a. lintel
    b. column
    c. simple beam
    d. girder

29. _____ fires are fires in ordinary type combustibles such as paper, wood, rubber, and plastics.
    a. Class A
    b. Class B
    c. Class C
    d. Class D

30. Scene safety is _____.
    a. the responsibility of the ambulance
    b. the responsibility of fire
    c. the responsibility of law enforcement
    d. a product of a good assessment and is an ongoing process

31. A material with a specific gravity of 2.5 will _____ in water.
    a. float
    b. sink
    c. mix evenly
    d. evaporate

32. Which is not a way in which water is supplied?
    a. Gravity feed
    b. Pumped system
    c. Combination gravity-pumped system
    d. Direct gravity-pumped system

33. Foam usually uses _____ to create a blanket over the surface of the fuel to cool and smother the fire, while sealing the escaping vapors.
    a. air
    b. carbon monoxide
    c. O2
    d. argon

34. _____ is the deterioration of concrete by the loss of surface material due to the expansion of moisture when exposed to heat.
    a. Torsion
    b. Spalling
    c. Impacting
    d. Veneer

35. In addition to the orange response guide section, a highlighted chemical in the yellow or blue section requires the reader to look in which section?
    a. Response guides
    b. Table of initial isolation and evacuations table
    c. Placard guides
    d. Truck and railcar guides

36. The _____ is useful when an "end of the line" knot is needed.
    a. basic figure-eight knot
    b. safety
    c. bowline
    d. Becket bend

37. Work _____ is the effort and physical training put forth to better perform physical tasks without overstressing or injuring the individual.
    a. strengthening
    b. training
    c. education
    d. hardening

38. If a hose bursts during fire attack, and there is no clamp available, _____.
    a. a firefighter has no choice but to have the pump operator shut down the line
    b. the firefighter can fold the hose twice over itself and kneel down to hold pressure buildup in the kinks
    c. the firefighter can use an airpack to clamp off the line
    d. the hose can still be used, provided it is supplying sufficient psi at the nozzle

39. _____ are butterfly valves that are opened and closed to control water flow.
    a. Gate valves
    b. Check valves
    c. Backflow valves
    d. PS&Y

40. Which chemical causes the majority of fatalities in chemical accidents?
    a. Kerosene
    b. Chlorine
    c. Ammonia
    d. Gasoline

41. Which WMD materials use the term SLUDGEM to describe the exposure symptoms?
    a. Blister agents
    b. Nerve agents
    c. Nuclear agents
    d. Biological agents

42. A team of firefighters exits a structure after completing their assigned task. The team is told to report to the _____ area for rest and hydration.

    a. staging

    b. accountability

    c. rehabilitation

    d. command post

43. A standard wildland hose load is the modified _____.

    a. Gasner bar pack

    b. horseshoe pack

    c. reverse-drain and carry pack

    d. shoulder loop

44. What type of ladder is used to provide secure footing for firefighters operating on a sloped roof?

    a. A folding or attic ladder

    b. A roof or hook ladder

    c. An A-frame combination ladder

    d. An extension ladder

45. A _____ is used to secure the firefighter to the ladder when both hands must be used to perform a task and a ladder belt is not available.

    a. rung lock

    b. halyard

    c. leg lock

    d. carry lock

46. In Early American history, insurance companies marked the occupancies they protected with signs on sheets of metal telling the firefighters which company held the insurance policy on the building. What were these signs called?

    a. Firemarks

    b. Fire wardens

    c. Fire societies

    d. Insurance marks

47. A(n) _____ attitude can be dangerous and result in PPE shortcuts.

    a. double-dressing

    b. dress-down

    c. wait-and-see

    d. can-do

48. Firefighters should never enter a _____ environment without first engaging SCBA.

    a. total flooding

    b. deluge

    c. wet pipe

    d. none of the above

49. When a victim is in moderate or fast-moving water, a firefighter should not use _____.

    a. a tether line

    b. PFD

    c. an additional PFD

    d. none of the above

50. The progressive end of a ground cover or wildland fire is known as the _____.

    a. head

    b. tail

    c. arrow

    d. flank

51. A flathead axe and a Halligan tool form a set known as the _____.

    a. kit

    b. irons

    c. force pack

    d. all of the above

52. Which form of transportation uses STCC codes?

    a. Highway

    b. Rail

    c. Air

    d. Water

53. The LEL for methane is 5%. When using a flammable gas detector calibrated for methane, the meter indicates a level of 100%. What is the amount of methane present by volume?

    a. 100%

    b. 50%

    c. 25%

    d. 5%

54. The mode of transportation that uses a dangerous cargo manifest is _____.

    a. highway

    b. rail

    c. water

    d. air

55. _____ occurs when the gases produced by fire stratify into layers based on their temperatures.
    a. Thermal layering
    b. Rollover
    c. Flashover
    d. Convection

56. A _____ wall is the extension of a wall past the top of a roof and can present a collapse hazard at the fire scene.
    a. mansard
    b. gambrel
    c. dormer
    d. parapet

57. When cutting into a roof, a _____ will help support the underlying Q-decking.
    a. trench cut
    b. triangle cut
    c. center rafter cut
    d. louver

58. Which agency primarily issues rules governing worker safety?
    a. EPA
    b. OSHA
    c. DOT
    d. DOD

59. The simplest–and often quickest–way to stop water flow from an individual sprinkler head is to _____.
    a. insert a stop
    b. turn off the valve head
    c. reinsert a fusible link
    d. find the water main

60. The two types of SCBA are _____.
    a. Scott and MSA
    b. backpack and compact packs
    c. hoop-wrap and steel cylinder
    d. open and closed circuit

61. "Any substance or material in any form or quantity that poses an unreasonable risk to safety, health, and property when transported in commerce" is the definition for a hazardous material which is used by which agency?
    a. EPA
    b. OSHA
    c. MSDS
    d. DOT

62. Firefighters should make every effort to assist in preserving _____ that may be important to fire investigators.

    a. point of origin

    b. smoke damage

    c. evidence

    d. depth of char

63. When a firefighter is performing a task that was not assigned to him or performing a task alone, he is said to be _____.

    a. accountable

    b. freelancing

    c. performing as trained

    d. fit for duty

64. A system that completely fills the area with an extinguishing agent to smother or cool the fire, or break the chain reaction is called a(n) _____ system. This system uses carbon dioxide or other inert gases, halogenated or clean agents, dry chemicals, or foam as extinguishing agents.

    a. deluge

    b. residential

    c. total flooding

    d. none of the above

65. Before performing decontamination on humans, who should be a priority to consult with?

    a. Trucking company

    b. Police officer

    c. Chemist

    d. OSHA

66. Who is responsible for clearing the hose bed prior to charging any of the lines?

    a. The officer

    b. The nozzleman

    c. The driver

    d. The IC

67. The use of SCBA prevents which route of exposure?

    a. Absorption

    b. Inhalation

    c. Injection

    d. Headache

68. The chainsaw can be used to cut through _____ metal.

    a. heavy-gauge

    b. light-gauge

    c. medium-gauge

    d. none of the above

69. When hoisting small equipment, it is recommended to use a _____.

    a. water knot

    b. tag line

    c. bowline

    d. granny knot

70. The command _____ is used to tell firefighters to turn a ladder right or left.

    a. raise

    b. retract

    c. extend

    d. rotate

71. What is the form required to list chemical hazards at a workplace?

    a. EHS Tier 3

    b. MSDS

    c. HazWoper

    d. DNRS

72. Simply defined, vehicle stabilization involves _____.

    a. chocking the tires

    b. deflating the tires

    c. taking weight off the vehicle's suspension system

    d. all of the above

73. Which is the movement of chemicals on the molecular level through PPE?

    a. Penetration

    b. Degradation

    c. Permeation

    d. Disintegration

74. Installed fire-extinguishing systems _____.

    a. allow SCBA use to be discontinued

    b. can create an oxygen-deficient atmosphere

    c. extinguish the fire and do not require firefighters to fight fire offensively

    d. require SCBA to have a low-pressure breathing hose

75. _____ fires are fires in combustible cooking fuels such as vegetable and animal oils and fats.
    a. Class B
    b. Class C
    c. Class D
    d. Class K

76. Pre-piped water lines for firefighter fire streams in a building are known as _____ systems.
    a. sprinkler
    b. standpipe
    c. hydrant
    d. exterior water supply

77. A minimum collapse zone should be _____ the height of the building.
    a. one-half
    b. equal to
    c. 1.5 times
    d. twice

78. The first step in filling an SCBA cylinder is _____.
    a. placing the cylinder in the fragmentation containment device
    b. connecting the fill hose to the cylinder
    c. opening the cylinder valve
    d. inspecting for hydro date

79. _____ fires are fires in combustibles metals, such as magnesium.
    a. Class A
    b. Class B
    c. Class C
    d. Class D

80. SCBA must be maintained _____.
    a. daily
    b. monthly
    c. annually
    d. all of these

81. The Tier 2 chemical reporting form is for which time period?
    a. What is on site currently
    b. The previous year
    c. The next year
    d. When the largest amount is present

82. What protective action is not recommend for radiation emergencies?

    a. Time

    b. Distance

    c. Shielding

    d. Oxygen sampling

83. PPV uses a technique that _____.

    a. sucks smoke

    b. pressurizes smoke

    c. replaces smoke with water

    d. removes smoke with water

84. The _____ section chief is responsible for documenting cost of materials and personnel for the incident.

    a. operations

    b. logistics

    c. planning

    d. finance/administration

85. What is the risk when the oxygen exceeds 23.5%?

    a. Fire

    b. Deficiency

    c. Toxicity

    d. Cancer

86. In the absence of power tools, breeching a masonry wall begins _____.

    a. with a large blow delivered by a flat head axe

    b. with a large blow to the brick, followed by prying on the brick

    c. by removing mortar around the brick

    d. shaving the face of the brick to find the weak point and then applying a forceful blow

87. A _____ is the designation for a set number of resources of the same type and kind.

    a. crew

    b. strike team

    c. type

    d. section

88. Which of the following can be a sign of a potential building collapse during a fire?

    a. Deterioration of mortar joints and masonry

    b. Sagging floors

    c. Smoke coming from cracks in walls

    d. All of the above

89. The most common type of construction for natural fiber ropes is _____.

    a. braided

    b. braid-on-braid

    c. kernmantel

    d. laid

90. A yellow star is the label for a fire extinguisher for _____.

    a. ordinary combustibles

    b. flammable and combustible liquids, gases, and greases

    c. energized electrical equipment

    d. combustible metals and alloys

91. DOT Table 1 materials require placards at what level?

    a. 1,001 pounds

    b. 1,001 liters

    c. Any amount

    d. Only with other hazardous materials

92. Foam equals _____.

    a. gas bubbles

    b. heavy water solutions of only animal proteins

    c. poor nozzle control results

    d. H20+F2(1+NACL)

93. In the NFPA 704 system, what number indicates the maximum hazard?

    a. 0

    b. 1

    c. 2

    d. 4

94. Seat-mounted SCBA _____.

    a. allows a firefighter to don the SCBA en route to an emergency

    b. allows for less cylinder failures due to the storage technique

    c. interferes with bunker gear donning en route to a call

    d. can interfere with the cylinder gauge check

95. Batteries in smoke detectors should be checked monthly and replaced every _____ months.

    a. three

    b. six

    c. nine

    d. twelve

96. Class C extinguisher _____.
    a. testing checks the conductivity of the agent and the nozzle
    b. fire tests determine the extinguisher rating
    c. numbers are based on the rating of the chemicals
    d. should be used as a last resort

97. _____ occurs when the entire contents of a compartment ignite almost simultaneously, generating intense heat and flames.
    a. Thermal layering
    b. Rollover
    c. Flashover
    d. Backdraft

98. A(n) _____ is a scrape or brush of the skin usually making it reddish in color and resulting in minor capillary bleeding.
    a. avulsion
    b. abrasion
    c. incision
    d. laceration

99. The _____ moves the coupling to another point in the load.
    a. accordion load
    b. Dutchman
    c. shoulder loop
    d. street drag

100. If an SCBA is stored in a compartment on the apparatus, the firefighter may have to use the _____.
    a. coat method
    b. sweatshirt method
    c. buddy donning method
    d. SCBA from another engine

Phase I, Exam III: Answers to Questions

1. F	26. C	51. B	76. B
2. F	27. B	52. B	77. C
3. T	28. D	53. D	78. D
4. T	29. A	54. C	79. D
5. F	30. D	55. A	80. D
6. T	31. B	56. D	81. B
7. T	32. D	57. B	82. D
8. F	33. A	58. B	83. B
9. T	34. B	59. A	84. D
10. F	35. B	60. D	85. A
11. T	36. A	61. D	86. C
12. T	37. D	62. C	87. B
13. F	38. B	63. B	88. D
14. T	39. A	64. C	89. D
15. F	40. D	65. C	90. D
16. F	41. B	66. C	91. C
17. F	42. C	67. B	92. A
18. F	43. A	68. B	93. D
19. F	44. B	69. B	94. D
20. T	45. C	70. D	95. B
21. C	46. A	71. B	96. A
22. C	47. C	72. C	97. C
23. A	48. A	73. C	98. B
24. C	49. A	74. B	99. B
25. B	50. A	75. D	100. A

Phase I, Exam III:
Rationale & References for Questions

Question #1. Complete and accurate communications center records should be maintained on all responses. NFPA 1001: 5.2.3. *FFHB, 2E:* Page 68.

Question #2. Firefighters and emergency medical providers respond to incidents that are often immediately dangerous to life and health. The term for this is IDLH. NFPA 1001: 5.1.1.2. *FFHB, 2E:* Page 125.

Question #3. When climbing a ladder, firefighters should keep their eyes focused level and ahead, occasionally looking up the climbing path for hazards and to view the objective. NFPA 1001: 5.36; 5.3.9; 5.3.10; 5.3.11; 5.3.12. *FFHB, 2E:* Page 383.

Question #4. Depending on the manufacturer's style, the age of the unit and the type of cylinder used, SCBA units will add 23 to 35 pounds of weight and 9 to 15 inches to the profile of the firefighter. NFPA 1001: 5.3.1. *FFHB, 2E:* Page 150.

Question #5. Hoselines can be hoisted whether charged or uncharged. NFPA 1001: 5.1.1.1. *FFHB, 2E:* Page 451.

Question #6. During fire operations, the potential for structural failure during a fire always exists. NFPA 1001: 5.3.10; 6.3.2. *FFHB, 2E:* Page 365.

Question #7. It is important to note, however, that PPE provides a minimum level of protection and should be considered the last resort of protection for firefighters and emergency responders operating at an incident. NFPA 1001: 5.1.1.2. *FFHB, 2E:* Page 125.

Question #8. For fire department operations, there are three standard methods of forcible entry:1.Conventional 2.Through-the-lock 3.Power tools NFPA 1001: 5.3.4; 6.3.2. *FFHB, 2E:* Page 530.

Question #9. Firefighters should avoid touching any radio antenna during transmission to avoid burns that can result from radio-frequency energy. NFPA 1001: 5.2.3. *FFHB, 2E:* Page 65.

Question #10. For equipment to be safe, it must be inspected and maintained. NFPA 1001: 5.5.3; 5.5.4; 6.1.1.1; 6.5.2. *FFHB, 2E:* Page 113.

Question #11. This exposes the lock mechanism, which is operated with the proper key tool. NFPA 1001: 5.3.4. *FFHB, 2E:* Page 519.

Question #12. Total flooding systems are used to protect an entire area, room, or building. NFPA 1001: 6.5.1. *FFHB, 2E:* Page 332.

Question #13. The wrong extinguisher can be worse than no extinguisher. NFPA 1001: 5.3.16. *FFHB, 2E:* Page 187.

Question #14. According to NFPA 1901, Standard for Automotive Fire Apparatus, aerial fire apparatus must have a minimum of 115 feet of ground ladders, including at least one attic ladder, two straightladders with folding roof hooks, and two extension ladders. NFPA 1001: 5.1.1.1. *FFHB, 2E:* Page 29.

Question #15. It is suggested that even after disconnecting the power to air bags, caution should be used when working around them. NFPA 1001: 5.3.7. *FFHB, 2E:* Page 600.

Question #16. Safety The water knot is the only knot that is recommended for use when tying webbing. NFPA 1001: 5.1.1.1. *FFHB, 2E:* Page 439.

Question #17. Water dictates the fire flow capacity or the amount of water that can be flowed. NFPA 1001: 5.3.15. *FFHB, 2E:* Page 203.

Question #18. Knockdown speed is how fast the foam spreads across the surface of a fuel. NFPA 1001: 5.3 NFPA 6.3. *FFHB, 2E:* Page 296.

Question #19. Advancing a hoseline up a ladder can be done with the line charged or uncharged. NFPA 1001: 5.3.8. *FFHB, 2E:* Page 257.

Question #20. This type of collapse usually results in a significant void being created near the remaining wall. NFPA 1001: 6.4.2. *FFHB, 2E:* Page 495.

Question #21. Understanding protective clothing and its relationship to toxicology and health effects is important NFPA 472: 5.3.3 and 5.4.3. *FFHB, 2E:* Page 845.

Question #22. Safety It is not only inappropriate, it is dangerous and contrary to NFPA 1983, Standard on Fire Service Life Safety Rope and System Components, to use the same rope for both utility and life safety operations. NFPA 1001: 5.1.1.1. *FFHB, 2E:* Page 424.

Question #23. A fire investigator will often look for a V pattern to help discover the point of origin of a fire. NFPA 1001: 6.3.4. *FFHB, 2E:* Page 654.

Question #24. Responders should understand the protective actions that are available to utilize. NFPA 472: 5.3.2 - 5.3.4. *FFHB, 2E:* Page 901.

Question #25. The benefits of self-discipline applied to PPE completeness pay a dividend in the form of acclimation. NFPA 1001: 5.1.1.2. *FFHB, 2E:* Page 137.

Question #26. The National Fire Protection Association (NFPA) created a 1500 series of standards that specifically focuses on safety issues. NFPA 1001: 6.1.1.1. *FFHB, 2E:* Page 110.

Question #27. The number rating compares approximately to the square footage to be extinguished. NFPA 1001: 5.3.16. *FFHB, 2E:* Page 194.

Question #28. A girder is a beam that supports other beams. NFPA 1001: 5.3.10; 6.3.2. *FFHB, 2E:* Page 346.

Question #29. Class A fires are fires in ordinary type combustibles such as paper, wood, rubber, and plastics. NFPA 1001: 5.3; 6.3. *FFHB, 2E:* Page 100.

Question #30. Good scene assessment considers many facts and probabilities. It is important to mention that the scene assessment is an ongoing process. NFPA 1001: 5.2.1. *FFHB, 2E:* Page 486.

Question #31. First responders should have an understanding of chemical and physical properties. NFPA 472: 5.2.2 - 5.2.4. *FFHB, 2E:* Page 798.

Question #32. Water is supplied in three ways. The first is gravity fed. The next is a pumped system. The third type is a combination gravity-pumped system. NFPA 1001: 5.3.15. *FFHB, 2E:* Page 205.

Question #33. The bubbles are filled with a gas, usually air, creating a blanket over the surface of the fuel to cool and smother the fire, while sealing the escaping vapors. NFPA 1001: 5.3 NFPA 6.3. *FFHB, 2E:* Page 294.

Question #34. Spalling is the deterioration of concrete by the loss of surface material due to the expansion of moisture when exposed to heat. NFPA 1001: 5.3.10; 6.3.2. *FFHB, 2E:* Page 350.

Question #35. Specific chemical information is important for a safe and informed response. NFPA 472: 4.2.1 - 4.2.2. *FFHB, 2E:* Page 812.

Question #36. The basic figure-eight knot is useful when an "end of the line" knot is needed, such as when a rappel rope is not long enough to reach the ground or a solid landing. NFPA 1001: 5.1.1.1. *FFHB, 2E:* Page 433.

Question #37. Work hardening is the effort and physical training put forth to better perform physical tasks without overstressing or injuring the individual. NFPA 1001: 6.1.1.1. *FFHB, 2E:* Page 115.

Question #38. If no clamp is available, a firefighter can fold the hose twice over itself and kneel down to hold pressure buildup in the kinks. NFPA 1001: 5.3.10. *FFHB, 2E:* Page 269.

Question #39. The gate valves are butterfly valves that are opened and closed to control water flow. NFPA 1001: 5.3.15. *FFHB, 2E:* Page 209.

Question #40. Understanding various protective actions is important for responder health and safety. NFPA 472: 5.2.3; 5.2.4. *FFHB, 2E:* Page 871.

Question #41. First responders should have a basic awareness of the threat of terrorism and basic response actions. NFPA 472: 4.2.1. *FFHB, 2E:* Page 928.

Question #42. The rehabilitation area is where firefighters can rest and rehydrate after performing fireground activities. NFPA 1001:5.3; 6.3. *FFHB, 2E:* Page 730-731.

Question #43. A standard wildland hose load is the modified Gasner bar pack, which provides ease of rolling and stretching the line, convenience of carrying with hands free, and protection for the couplings. NFPA 1001: 5.1.1.1. *FFHB, 2E:* Page 250.

Question #44. A roof or hook ladder is used to provide secure footing for firefighters operating on a sloped roof. NFPA 1001: 5.36; 5.3.9; 5.3.10; 5.3.11; 5.3.12. *FFHB, 2E:* Pages 378-379.

Question #45. A leg lock is used to secure the firefighter to the ladder when both hands must be used to perform a task and a ladder belt is not available. NFPA 1001: 5.36; 5.3.9; 5.3.10; 5.3.11; 5.3.12. *FFHB, 2E:* Page 408.

Question #46. In early American history, insurance companies marked the occupancies they protected with signs on sheets of metal, known as firemarks, telling the firefighters which company held the insurance policy on the building. NFPA 1001: N/A. *FFHB, 2E:* Page 8.

Question #47. Firefighters who take a wait-and-see attitude to decide what level of PPE they are going to need, have set themselves up to shortcut their PPE ensemble if, on arrival at the scene, the situation requires immediate life-saving actions. NFPA 1001: 5.1.1.2. *FFHB, 2E:* Page 137.

Question #48. Firefighters must never enter a total flooding environment without first engaging their SCBA. NFPA 1001: 6.5.1. *FFHB, 2E:* Page 336.

Question #49. In still or extremely slow-moving water, it is a good idea for the rescuer to swim out holding a tether line. A rescuer should not use an attached tether line in moderate to fast-moving water. It can pull the rescuer under. NFPA 1001: 6.4.2. *FFHB, 2E:* Page 493.

Question #50. From there, the team will work up one of the flanks of the fire working toward the head or progressive end of the fire. NFPA 1001: 5.3.19. *FFHB, 2E:* Page 621.

Question #51. Together, the axe and Halligan tool form the "irons." NFPA 1001: 5.3.4. *FFHB, 2E:* Page 512.

Question #52. Specific chemical information is important for a safe and informed response. NFPA 472: 4.2.1 - 4.2.2. *FFHB, 2E:* Page 825.

Question #53. Responders should understand the protective actions that are available to utilize. NFPA 472: 6.2.1 - 6.2.2. *FFHB, 2E:* Page 901.

Question #54. Specific chemical information is important for a safe and informed response. NFPA 472: 4.2.1 - 4.2.2. *FFHB, 2E:* Page 825.

Question #55. Thermal layering occurs when the gases produced by fire stratify into layers based on their temperatures. NFPA 1001: 5.3; 6.3. *FFHB, 2E:* Page 99.

Question #56. A parapet wall is the extension of a wall past the top of a roof and can present a collapse hazard at the fire scene. NFPA 1001: 5.3.10; 6.3.2. *FFHB, 2E:* Page 363.

Question #57. The use of a triangular cut will help support the underlying Q-decking because it is interlocked. NFPA 1001: 5.3.11. *FFHB, 2E:* Page 579.

Question #58. Responders should understand the rules that govern their activities. NFPA 472: 4.2.1. *FFHB, 2E:* Page 744.

Question #59. The simplest–and often quickest–way to stop water flow from an individual sprinkler head is to insert a stop. NFPA 1001: 6.5.1. *FFHB, 2E:* Page 326.

Question #60. Two different types of SCBA in general use in today's fire service are open-circuit and closed-circuit systems. NFPA 1001: 5.3.1. *FFHB, 2E:* Page 151.

Question #61. Understanding the regulations covering hazardous materials is important. NFPA 472: 4.2.1. *FFHB, 2E:* Page 744.

Question #62. Firefighters should make every effort to assist in preserving evidence that may be important to fire investigators. NFPA 1001: 6.3.4. *FFHB, 2E:* Page 653.

Question #63. When a firefighter is performing a task that was not assigned to him or performing a task alone, he is said to be freelancing. NFPA 1001:5.3; 6.3. *FFHB, 2E:* Page 725.

Question #64. Total flooding systems are used to protect an entire area, room, or building. The total flooding system discharges an extinguishing agent that completely fills or floods the area with the extinguishing agent to smother or cool the fire or break the chain reaction. This system uses carbon dioxide or other inert gases, halogenated or clean agents, dry chemicals, or foam as extinguishing agents. NFPA 1001: 6.5.1. *FFHB, 2E:* Page 332.

Question #65. Understanding various protective actions is important for responder health and safety. NFPA 472:5.3.4-5.4.1. *FFHB, 2E:* Page 877.

Question #66. The driver is responsible for clearing the hose bed prior to charging any of the lines. NFPA 1001: 5.3. *FFHB, 2E:* Page 251.

Question #67. Understanding protective clothing and its relationship to toxicology and health effects is important. NFPA 472: 5.3.3; 5.4.3. *FFHB, 2E:* Page 835.

Question #68. The chainsaw can also be used to cut through wood siding, woodframe walls, certain doors, and light-gauge metal. NFPA 1001: 5.3.4. *FFHB, 2E:* Page 516.

Question #69. NFPA 1001: 5.1.1.1. *FFHB, 2E:* Page 451.

Question #70. The command rotate is used to tell firefighters to turn a ladder right or left. NFPA 1001: 5.36; 5.3.9; 5.3.10; 5.3.11; 5.3.12. *FFHB, 2E:* Page 375.

Question #71. Material safety data sheets (MSDS) provide safety and chemical information. NFPA 472: 4.2.1-4.2.3. *FFHB, 2E:* Page 342.

Question #72. In any case, if a vehicle has an injured person inside, it must be stabilized by taking the weight off the vehicle's suspension system.(Note: Deflating the tires will not accomplish this.) NFPA 1001: 5.2.1. *FFHB, 2E:* Page 487.

Question #73. Understanding protective clothing and its relationship to toxicology and health effects is important. NFPA 472: 5.3.3; 5.4.3. *FFHB, 2E:* Page 842.

Question #74. Installed fire-extinguishing systems such as total flooding carbon dioxide or halon systems create an oxygen-deficient atmosphere. NFPA 1001: 5.3.1. *FFHB, 2E:* Page 145.

Question #75. Class K fires are fires in combustible cooking fuels such as vegetable and animal oils and fats. NFPA 1001: 5.3; 6.3. *FFHB, 2E:* Page 100.

Question #76. Standpipe systems are designed to allow firefighters to fight fires in larger buildings by pre-piping water lines for fire streams throughout the building. NFPA 1001: 6.5.1. *FFHB, 2E:* Page 328.

Question #77. A minimum collapse zone should be 1.5 times the height of the building. NFPA 1001: 5.3.10; 6.3.2. *FFHB, 2E:* Page 365.

Question #78. 1.Check the hydrostatic test date of the cylinder, JPR 7-8A. NFPA 1001: 5.3.1. *FFHB, 2E:* Page 177.

Question #79. Class D fires are fires in combustibles metals, such as magnesium. NFPA 1001: 5.3; 6.3. *FFHB, 2E:* Page 100.

Question #80. Daily Maintenance SCBA units should be checked daily to ensure they are secured and ready for operation.Monthly Maintenance The monthly SCBA check contains all elements of the daily check, but adds several checks of the mechanics of the system.Annual and Bi-annual Maintenance NIOSH and SCBA manufacturers require a number of different functional tests of SCBA units NFPA 1001: 5.3.1. *FFHB, 2E:* Page 172.

Question #81. Responders should have a basic understanding of the SARA regulation. NFPA 472: 4.2.1. *FFHB, 2E:* Page 746.

Question #82. Understanding various protective actions is important for responder health and safety. NFPA 472: 5.2.2 - 5.2.4. *FFHB, 2E:* Page 875.

Question #83. The positive-pressure technique actually injects air into the compartment and pressurizes it. NFPA 1001: 5.3.11. *FFHB, 2E:* Page 573.

Question #84. The finance/administration section chief is responsible for documenting cost of materials and personnel for the incident. NFPA 1001: 6.1.1.1. *FFHB, 2E:* Page 40.

Question #85. Responders should understand the protective actions that are available to utilize. NFPA 472: 5.3.2 - 5.3.4. *FFHB, 2E:* Page 901.

Question #86. Work at the mortar joints because this is usually the weak point. NFPA 1001: 5.3.4; 6.3.2. *FFHB, 2E:* Page 544.

Question #87. A strike team is the designation for a set number of resources of the same type and kind. NFPA 1001: 6.1.1.1. *FFHB, 2E:* Page 42.

Question #88. There are numerous signs that can indicate a possible collapse in a structure. All of these are included as potential signs. NFPA 1001: 5.3.10; 6.3.2. *FFHB, 2E:* Page 364.

Question #89. As previously mentioned, the laid method is the most common type of construction for natural fiber ropes. NFPA 1001: 5.1.1.1. *FFHB, 2E:* Page 422.

Question #90. Figure 8-3 NFPA 1001: 5.3.16. *FFHB, 2E:* Page 189.

Question #91. Responders should understand basic recognition and identification skills. NFPA 472: 5.2.2 - 5.2.4. *FFHB, 2E:* Page 758.

Question #92. Foam is an aggregate of gas-filled bubbles formed from aqueous solutions of specially formulated concentrated liquid foaming agents. NFPA 1001: 5.3 NFPA 6.3. *FFHB, 2E:* Page 294.

Question #93. Responders should understand basic recognition and identification skills. NFPA 472: 5.2.2 - 5.2.4. *FFHB, 2E:* Page 768.

Question #94. From the seated position, it is extremely difficult, if not impossible, to check the cylinder gauge and compare it to the regulator gauge. NFPA 1001: 5.3.1. *FFHB, 2E:* Page 163.

Question #95. Batteries in smoke detectors should be checked monthly and replaced every six months. NFPA 1001: 5.5.1. *FFHB, 2E:* Page 679.

Question #96. The testing of Class C extinguishers and agents tests only the conductivity of the agent and the nozzle, or hose and nozzle combination. NFPA 1001: 5.3.16. *FFHB, 2E:* Page 194.

Question #97. Flashover occurs when the entire contents of a compartment ignite almost simultaneously, generating intense heat and flames. NFPA 1001: 5.3; 6.3. *FFHB, 2E:* Page 99.

Question #98. An abrasion is a scrape or brush of the skin usually making it reddish in color and resulting in minor capillary bleeding. NFPA 1001: 4.3. *FFHB, 2E:* Page 712.

Question #99. The dutchman moves the coupling to another point in the load. NFPA 1001: 5.1.1.1. *FFHB, 2E:* Page 236.

Question #100. If the cabinet height or mounting bracket position off the SCBA does not allow for ease of donning while standing, the firefighter should remove the SCBA unit and use the "coat" or "over-the-head" method of donning. NFPA 1001: 5.3.1. *FFHB, 2E:* Page 164.

PHASE II

APPLICATION & ANALYSIS

Section two is evaluating for a higher level of learning. Within this section, we are testing to determine an understanding of comparing material, describing processes, explaining procedures, and interpreting results. A test-taker mastering this section should have a better grasp of the material and a greater depth of understanding. Referring to Table I-1 (Bloom's Taxonomy, Cognitive Domain), we are addressing the following levels:

- application
- analysis

1. Class C fires involve any electrical equipment.

    a. True

    b. False

2. For a fire involving a large amount of shredded paper, a pressurized water extinguisher is acceptable.

    a. True

    b. False

3. When fighting an interior structure fire, it is important to remember that the temperature of smoke can be as or more hazardous than the smoke itself.

    a. True

    b. False

4. You are doing a search of a smoke-filled retirement home and you find an older, very overweight man unconscious and lying on top of his bed. It is acceptable for you and your partner to use the bedspread to drag him off the bed and to safety.

    a. True

    b. False

5. Overhaul operations are complete at a structure fire and there were no additional signs of hidden fires found by the firefighters; therefore, there is no need to revisit the structure.

    a. True

    b. False

6. When conducting salvage operations, firefighters can help ensure their safety by knowing what other operations are ongoing and where they are taking place.

    a. True

    b. False

7. It is imperative for all firefighters to be familiar with some sort of rescue knot.

    a. True

    b. False

8. You are given the assignment to advance a 1-3/4 line to the fire floor (second floor) of an apartment fire. The fire is room and contents in a bedroom. It would be acceptable to advance the hoseline dry and call for it to be charged upon reaching the apartment.

    a. True

    b. False

9. The invention of the triple combination engine company was a great innovation in the fire service because it carried water, hoses, and an aerial ladder device.

   a. True

   b. False

10. During a wildland fire operation, your engine company is sent to defend three homes in an affluent area of the city. If you are forced to make a stand for these homes, it is important to remember that you may have an alternate source of water in the backyard.

    a. True

    b. False

11. Head pressure measures the pressure at the top of a column of water.

    a. True

    b. False

12. As matter interacts, substances are formed, changed, and destroyed. Matter doesn't disappear; it merely changes form.

    a. True

    b. False

13. Fire department standard operating procedures cover topics or issues that rarely change and, therefore, SOPs never need revision.

    a. True

    b. False

14. As you are fighting a propane tank fire, you should remember that the best strategy might be to cool the tank and let the product burn.

    a. True

    b. False

15. When connecting to a standpipe inside a structure, you must have faith that the driver/operator will complete his portion of the evolution.

    a. True

    b. False

16. When responding to a fire protective system activation, firefighters should go immediately to the annunciator panel for information about the activation.

    a. True

    b. False

17. The advantage of fighting a ground cover fire in light fuels is that firefighting operations are much safer.

    a. True

    b. False

18. While inspecting cooking surfaces and hood and duct systems, firefighters should ensure that the extinguishing agent cylinders are charged and armed and the manual activation control is hidden from view to prevent accidental activations.

    a. True

    b. False

19. You arrive on scene of a residential structure fire. Upon arrival on scene, you see very dark smoke puffing in and out under the front door. You suspect a flashover situation is possible.

    a. True

    b. False

20. Procedures can be classified as formal or informal. An example of a formal procedure is a company officer's instructions at the beginning of a shift. An example of an informal procedure is on-the-job training a new member receives.

    a. True

    b. False

21. Setting a knot is the practice of making sure all parts of the knot are lying in the properorientation to the other parts and look exactly as the pictures indicated.

    a. True

    b. False

22. On the scene of a small structure fire with sprinkler activation, the reporting party tells you that when the sprinkler head discharged, there was quite a bit of air in the line. You should not be concerned about this information..

    a. True

    b. False

23. The responsibility for firefighter safety is shared by two areas: the department itself (administration) and the individual firefighter.

    a. True

    b. False

24. After use on the fireground, salvage and overhaul tools and equipment do not need to be inspected, cleaned, and maintained like other firefighting equipment.

    a. True

    b. False

25. By-products of combustion can include water vapor, carbon particles, carbon monoxide, sulfur dioxide, and hydrogen cyanide.

    a. True

    b. False

26. The best method for drying fire service rope is to use a front loading dryer on high heat.

    a. True

    b. False

27. While fighting a fire with a 2½" line, you open the bale of the nozzle and feel like you are going to fall over backward. This is a classic example of nozzle reaction.

   a. True

   b. False

28. Your partner is a 110-lb. firefighter. She is attempting to use a 16-lb. sledgehammer to breach a wall in training. This is probably not the correct tool for her to use.

   a. True

   b. False

29. You find it necessary to tie a life line around a large metal light pole. The knot suitable for this operation would be a _____.

   a. basic figure-eight knot

   b. follow-through figure-eight knot

   c. bowline knot

   d. follow-through bowline knot

30. Firefighters are responding to a fire in a structure that is Type V construction. The most common occupancy of this type of construction is _____.

   a. residential

   b. commercial

   c. business

   d. educational

31. Your engine has parked at the scene of a structure fire. The driver operator tells you she is going to use a reverse lay. She is referring to _____.

   a. a direct attack

   b. an indirect attack

   c. a master stream

   d. water supply

32. Which of the following is an organic material?

   a. Wood

   b. Iron

   c. Granite

   d. Quartz

33. One firefighter is preparing to use a salvage cover to protect a large, unbreakable item. Which is the best deployment method to use for this situation?

   a. Counter payoff deployment

   b. Balloon toss

   c. Shoulder toss

   d. Roll toss

34. A dangerous placard indicates _____.

    a. extreme danger

    b. mixed load of hazardous materials

    c. less than 1,000 pounds of a chemical

    d. water-reactive materials

35. You are planning a confined-space drill for your company officer. You are planning on firefighters using open-circuit SARs. For escape purposes, you should not plan on any firefighters advancing more than _____.

    a. 5 to 10 minutes into the confined space, depending on the length of the hose

    b. 5 to 10 minutes into the confined space, depending on the size of the escape unit SCBA

    c. 5 to 10 minutes into the confined space as drills are not an excuse to take risks

    d. 5 to 10 minutes into the confined space, as that is what is allowable by law

36. What type of fitness activities are intended to raise the heart and breathing rates in order to strengthen the heart/lung relationship and reduce the chance of overexertion on the scene of an incident?

    a. Cardiovascular conditioning

    b. Core strengthening

    c. Flexibility improvement

    d. Resistance training

37. If a material is released onto the highway and the weather report is calling for rain, which table would provide assistance in the DOT ERG?

    a. Placard table

    b. Truck table

    c. List of dangerous water-reactive materials table

    d. Isolation table

38. The safety triad consists of three key components? Which of the following is not a key component of the safety triad?

    a. Personnel

    b. Environment

    c. Procedures

    d. Equipment

39. NFPA 1901, Standard for Automotive Fire Apparatus, states that a pumper should have a permanently mounted fire pump with a capacity of at least _____ gallons per minute and that the pumper should carry no less than _____ gallons of water.

    a. 250; 250

    b. 500; 200

    c. 500; 300

    d. 750; 300

40. You are at a residence with light smoke showing. The front door has a window next to it. You break the window and unlock the door from the inside. However, you still cannot open the door. What is the most likely problem?

    a. Door swelling

    b. Additional locking devices

    c. Furniture

    d. Debris

41. When connecting an engine to a fire hydrant, you notice a cap is loose. You decide it would be good practice to _____.

    a. tighten that cap and use the other side to connect to your pump

    b. open all caps and flush the hydrant before connecting the hose to your pump

    c. use another hydrant

    d. continue using the hydrant as normal

42. During a residential structure fire operation, there is a collapse of the interior first floor. You remind your partner that victims may survive _____.

    a. in the voids

    b. in the basement

    c. in the spans

    d. on the first floor

43. You are an A shift firefighter. At morning pass down, B shift engine company neglects to inform you they used a pressurized-water extinguisher on a mattress fire. You discover this when you go to use the extinguisher on a small trash fire in an office building. This scenario tells you something important. What is it?

    a. B shift company should be written up.

    b. A shift company officer is neglectful.

    c. B shift company officer is neglectful.

    d. A shift morning truck checks were not thoroughly completed.

44. The maximum quantity of flammable and combustible liquids permitted to be stored inside a building _____.

    a. varies according to the occupancy

    b. is 1,000 gallons

    c. is 500 gallons

    d. is not a concern of firefighters conducting inspections

45. You and your partner are removing a patient from a two-vehicle, high-speed, heavy-damage motor vehicle accident. The passenger SRS device is deployed, but the driver's SRS device has not. While removing the passenger, you remind your partner that _____.

    a. the driver's SRS device is not a concern

    b. the driver's SRS device is still a concern

    c. the SRS devices all deploy simultaneously

    d. SRS devices and motor vehicle accidents are not related

46. What can be fatal for upward of 80% of the affected patients?

    a. Heat cramps

    b. Heat stroke

    c. Heat exhaustion

    d. Heat rash

47. As hose is vital to firefighter operations, it is imperative that you remember that hose testing should occur _____.

    a. annually

    b. bi-annually

    c. daily

    d. visually during reloading

48. You and your partner are manning a 2½" exposure line on a large residential structure fire. You are becoming fatigued from holding the hose, so you momentarily shut down the line and you and your partner loop the hose and sit on it. You find this hose is much easier to control. This is an example of the ground absorbing the energy from the _____.

    a. friction

    b. nozzle reaction

    c. vacuum

    d. head pressure

49. Firefighter Morton is giving instructions to the operator of an aerial ladder. She wants to give the command to move the ladder from a horizontal position to a vertical angle. What command should she use?

    a. Raise

    b. Extend

    c. Retract

    d. Rotate

50. Which of the following American leaders played an important role in the development of the American fire service?

    a. Benjamin Franklin

    b. George Washington

    c. Thomas Jefferson

    d. All of the above

51. You are fighting an outdoor propane tank fire with your two partners. There is another hoseline advancing with three more firefighters and you are working in tandem. As you approach, you notice the heat appears to be increasing. You should _____.

    a. retreat

    b. decrease your fog pattern

    c. increase your fog pattern

    d. wait for the other team to advance

52. Making equipment safe is addressed in three ways. Which of the following is not one of these three ways?
    a. Equipment selection
    b. Equipment application
    c. Equipment inspection and maintenance
    d. Equipment complexity

53. When initially ventilating vertically, sometimes _____ can be initiated before a roof cut is made.
    a. opening a dumbwaiter bulkhead
    b. opening a first floor door
    c. opening a fire floor window
    d. turning on the window air conditioner

54. As a rule of thumb, which materials will burn: organic or inorganic?
    a. Organic
    b. Inorganic
    c. Both are equally likely to burn.
    d. Neither are likely to burn.

55. Which is not one of the key factors in the FBI's definition of terrorism?
    a. The type of explosive
    b. Intimidation
    c. Violent act
    d. Political and social objectives

56. Although all PPE requires pre-hydration, which level is it most important?
    a. A
    b. B
    c. C
    d. D

57. As a firefighter, you know that anytime you encounter a PIV, WIV, or OS&Y, it should be _____.
    a. in the open position
    b. in the closed position
    c. locked in the open position
    d. locked in the closed position

58. Which training level can perform offensive operations at a chemical release?
    a. Awareness
    b. Operations
    c. Technician
    d. None

59. When fighting a wildland fire, it is generally much safer to be _____the fire.

    a. upslope from

    b. downslope from

    c. parallel to

    d. perpendicular to

60. Perhaps the most important task of using a non-integrated PASS device is ensuring _____.

    a. it can be heard outside

    b. it is turned on

    c. it is equipped with extra batteries

    d. the flashing light must be seen through smoke

61. You are fighting a shed fire in the back of a residential structure. Of the four conditions that present respiratory hazards, you are least concerned with _____.

    a. oxygen deficiency

    b. high temperatures

    c. smoke or other by-products of combustion

    d. toxic environments

62. A 4A rating of an extinguisher should extinguish _____.

    a. 4 square-feet of fire

    b. 4 times the fire of a 1A

    c. 16-square-feet of fire

    d. 4 lbs of class A material

63. What type of load is measured in Btus and is used mainly in fire engineering and not in the construction industry?

    a. Design load

    b. Concentrated load

    c. Impact load

    d. Fire load

64. Firefighters arrive on the scene to find an adult male patient who appears to be unconscious. The scene is safe and the firefighters are using BSI. They begin to assess the patient. What type of consent allows them to provide this care to the patient?

    a. Implied consent

    b. Express consent

    c. Unspoken consent

    d. Standard of care consent

65. Fires involving explosives should be _____.
   a. considered having minimal danger
   b. considered being very dangerous
   c. fought with two hose lines
   d. fought with two engines

66. At the hazardous materials operations level, when would the shutting of valves be appropriate?
   a. On rail cars
   b. When they are remote
   c. Only at fuel storage facilities
   d. Always

67. You are preparing to hoist an axe to the roof of a structure. Before you signal that the axe is ready to be hoisted, you tie a(n) _____ in the working end.
   a. overhand knot
   b. clove hitch
   c. Becket bend
   d. double Becket bend

68. What is a combustible gas detector designed to read?
   a. Lower flash points
   b. Lower explosive level
   c. Toxic and corrosive gases
   d. Upper flash point

69. Which construction method allows fire to spread to an attic more rapidly?
   a. Balloon frame construction
   b. Platform framing construction
   c. Both methods allow fire to spread to an attic at the same rate.
   d. Neither method allows for fire to spread to an attic.

70. You are fighting a fire in an industrial area of town at a business that produces pesticides. One of the tanks is burning freely. Your IC is gathering information and resources about the product. As a firefighter, you know that your strategy might be

   _____.
   a. an indirect attack
   b. a direct attack
   c. a transitional attack
   d. to let it burn

71. On the local level, the fire department should be involved in which planning group?
   a. LEPC
   b. SERC
   c. DDNR
   d. OSHA

72. What is combustion?

    a. A rapid, persistent chemical change that releases heat and light and is accompanied by flame

    b. Rapid oxidation with the development of heat and light

    c. A reaction that is a continuous combination of a fuel with certain elements, prominent among which is oxygen in either a free or combined form

    d. All of the above

73. You and your partner are advancing an attack line to a reported room and contents fire in an apartment. Upon reaching the door of the interior room where the fire is, you find that a couch at the far end of the room is burning freely, but there appears to be no real spread beyond the furniture. Using a(n) _____, you apply a fog pattern into the room and then close the door.

    a. aggressive straight stream attack

    b. direct attack

    c. combination attack

    d. indirect attack

74. You and your partner are on the nozzle of a 200-foot attack line that is uphill from the engine. You notice that the flow from the nozzle does not seem to be as strong as it normally is during drills. The most likely reason is _____.

    a. heat resistance

    b. nozzle reach

    c. nozzle reaction

    d. head pressure

75. You have been assigned the duty of stopping the flow from an activated sprinkler head. You should _____.

    a. rely on past practice to stop the flow and establish a leak-free seal

    b. use a wedge or sprinkler tongs

    c. prepare to get seriously wet

    d. all of the above

76. Which position is preferred?

    a. Uphill

    b. Uphill and downwind

    c. Upwind

    d. Downwind

77. Propane can be considered more dangerous than natural gas due to its

    _____.

    a. flammability

    b. vapor pressure

    c. vapor density

    d. ease of use

78. Photoionization (PID) detectors are designed to identify which risk?
    a. Corrosive
    b. Toxic
    c. Radioactive
    d. Explosive

79. Firefighter Johnson is preparing his personal protective equipment (PPE) at the beginning of his duty shift in the fire station. Which of the following is not something he should do to prepare his personal protective equipment?
    a. Make sure the PPE clothing material is dry.
    b. Pack all the PPE into the carrying bag to take it along on the apparatus.
    c. Check all the pocket tools in the PPE, like flashlights or webbing.
    d. Ensure that any alternative PPE items are available and ready for use, such as EMS or wildland gear.

80. You are setting up a positive pressure ventilation of a residence filled with smoke. The fire has been put out. You know that the quickest and easiest set up to prepare for overhaul is usually _____ ventilation.
    a. negative pressure
    b. positive pressure
    c. combination
    d. natural

81. You have responded to a call of a possible structure fire in a single-family residence. You arrive on scene and find smoke showing in a Spanish style home. The home has bars over the windows and security gates–all of which are locked. You decide your best option is to use a _____ for access.
    a. Bam Bam
    b. pick head axe
    c. bolt cutters
    d. rotary saw

82. Firefighter McDonald wants to give the command to the other firefighters on his team to increase the length and reach of the ladder with the fly sections. What command should he use?
    a. Raise
    b. Extend
    c. Retract
    d. Rotate

83. What is not an OSHA requirement to use a Level A garment?
    a. Above IDLH levels
    b. Toxic by skin absorption chemical
    c. Unidentified material
    d. None of the above

84. The NFPA states that "95 percent of alarms shall be answered within _____ seconds, and in no case shall the initial operator's response to an alarm exceed _____ seconds."
    a. 15; 30
    b. 30; 60
    c. 45; 60
    d. 60; 90

85. A firefighter is conducting a fire safety inspection in a public school. She can expect what kind of opening and latching devices to be on the exits?
    a. No-knowledge hardware
    b. Special egress control devices
    c. Panic hardware
    d. Interconnected hardware

86. You and your partner are assigned the task of advancing a hoseline to a second story window. Your partner takes an uncharged hoseline and begins ascent. For a safe operation, you should use _____ as a guide to begin your climb.
    a. the next coupling
    b. the middle of the next section
    c. the third loop
    d. the pre-marked section

87. Which book is an excellent reference for evacuation and isolation distances?
    a. NIOSH pocket guide
    b. MSDS
    c. DOT ERG
    d. They are not available.

88. Firefighters are fighting a fire in a building that was constructed with composite trusses made of both wood and steel. Why is this a dangerous situation for the firefighters?
    a. The mixture of steel and wood can cause rapid collapse because steel expands faster than wood, causing stress at the intersections of the two materials.
    b. The composite trusses have no tensile strength.
    c. The mixture of steel and wood can cause rapid collapse because wood expands faster than steel, causing stress at the intersections of the two materials.
    d. The composite trusses have no compressive strength.

89. You are explaining drafting to a rookie firefighter. She says that you are using an engine to suck the water out of a lake. You correct her and say that you are actually using _____.
    a. gravity and pumping
    b. check valves
    c. atmospheric pressure
    d. mechanical sucking

90. You are responding to a reported wildland fire incident. Before donning the wildland ensemble, you should ensure _____.
    a. fire-resistive or cotton undergarments are worn
    b. a bandana is tied around your neck
    c. boots are leather with a steel toe
    d. the fire shelter is open and ready to deploy

91. Which is the worst of the violations of the standard of care?
    a. Negligence
    b. Liability
    c. Gross negligence
    d. None of the above

92. You have been tasked with search and rescue on a residential structure fire. It is 3 A.M. The home is a two-story home. It can generally be assumed that your greatest rescue potential lies _____.
    a. on the uppermost floor
    b. in the basement
    c. on ground level
    d. both a and b

93. When first arriving at a chemical release, what is one of the important first steps?
    a. Protection of the responders
    b. Protection of those involved
    c. Preventing others from becoming involved
    d. All of the above

94. The basic goal of salvage operations is to _____.
    a. remove the harmful atmosphere from the material
    b. protect the material from the harmful atmosphere
    c. both a and b
    d. none of the above

95. If a responder is not intimately familiar with the DOT ERG, which page should they consult?
    a. Placard page
    b. How to use this guidebook page
    c. Hazard classification system page
    d. Water reactive table

96. Which of the following methods can be used to control external bleeding?
    a. Direct pressure on the site of bleeding
    b. Elevation of the bleeding injury
    c. Using pressure points
    d. All of the above

97. When responding to residential structure fires with sprinkler systems, it is important for firefighters to remember that these systems _____.

    a. are intended to protect vital areas, such as computer rooms or stereo centers

    b. are only intended to protect egress hallways

    c. are intended for life safety and not necessarily property conservation

    d. are intended for property conservation and not necessarily life safety

98. You are assisting the driver/operator of a pumper with a water supply operation. The driver/operator wants to draft water from a lake. After positioning the apparatus, she asks you to connect the supply line. You know that you will need to find the

    _____.

    a. large diameter supply line

    b. two 2-1/2-inch hoselines for drafting

    c. hard sleeve and strainer

    d. drafting manual

99. While on your way to a reported aircraft fire, you are going over your plan of action in your head. An advantage of the foam in your tank is that it

    _____.

    a. clings to vertical surfaces

    b. expands greatly

    c. does not require a special foam nozzle

    d. There are no advantages.

100. Firefighter Watkins is working in a busy emergency communications center. There are many phone lines ringing at once. Which of the following phone lines should Firefighter Watkins answer first?

    a. Emergency or 9-1-1 lines

    b. Direct lines

    c. Business lines

    d. Administrative lines

Phase II, Exam I: Answers to Questions

1. F	26. T	51. C	76. C
2. T	27. T	52. D	77. C
3. T	28. T	53. A	78. B
4. T	29. B	54. A	79. B
5. F	30. A	55. A	80. B
6. T	31. D	56. A	81. D
7. T	32. A	57. C	82. B
8. T	33. C	58. C	83. C
9. F	34. B	59. B	84. B
10. T	35. B	60. B	85. C
11. F	36. A	61. A	86. A
12. T	37. C	62. B	87. C
13. F	38. B	63. D	88. A
14. T	39. D	64. A	89. C
15. T	40. B	65. B	90. A
16. T	41. B	66. B	91. C
17. F	42. A	67. A	92. A
18. F	43. D	68. B	93. D
19. F	44. A	69. A	94. C
20. F	45. B	70. D	95. B
21. F	46. B	71. A	96. D
22. T	47. D	72. D	97. C
23. F	48. B	73. D	98. C
24. F	49. A	74. D	99. C
25. T	50. D	75. D	100. A

Phase II, Exam I:
Rationale & References for Questions

Question #1. Class C fires involve energized electrical equipment, which eliminates the use of water-based agents to extinguish them. NFPA 1001: 5.3.16. *FFHB, 2E:* Page 186.

Question #2. Class A fires involve ordinary combustibles such as wood, paper, cloth, plastics, and rubber.These fuels can be extinguished with water, water-based agents or foam, and multipurpose dry chemicals. NFPA 1001: 5.3.16. *FFHB, 2E:* Page 185.

Question #3. This combination of materials is an irritant to the respiratory system and, in many cases, small inhaled quantities may be fatal. In addition, the temperature of smoke may cause burns to the respiratory system. NFPA 1001: 5.3.1. *FFHB, 2E:* Page 146.

Question #4. Rescuers can move a patient by placing him or her on a blanket, bunker coat, salvage cover, and so on, or by using the patient's own clothing as a handhold or by the utilization of a webbing sling. NFPA 1001: 5.3.9. *FFHB, 2E:* Page 469.

Question #5. Even after all areas have been checked, it is important to revisit the scene of the incident. NFPA 1001: 5.3.13. *FFHB, 2E:* Page 650.

Question #6. When conducting salvage operations, firefighters can help ensure their safety by knowing what other operations are ongoing and where they are taking place. NFPA 1001: 5.3.14. *FFHB, 2E:* Page 642.

Question #7. However, it is also important to practice the rescue knot on other people to simulate victims needing rescue. An emergency scene is not the proper venue for learning skills. NFPA 1001: 5.1.1.1. *FFHB, 2E:* Page 437.

Question #8. If the fire does not involve the stairs, the best method is to advance an uncharged line to the fire floor. NFPA 1001: 5.3. *FFHB, 2E:* Page 254.

Question #9. The invention of the triple combination engine company was a great innovation in the fire service because it carried water, could pump the water, and carried hoses and other equipment for delivering the water to a fire. NFPA 1001: N/A. *FFHB, 2E:* Page 16.

Question #10. Firefighters should also be aware of other sources of water available to them such as swimming pools and any other body of water that can be accessed for firefighting. NFPA 1001: 5.3.15. *FFHB, 2E:* Page 205.

Question #11. Head pressure measures the pressure at the bottom of a column of water in feet. Head pressure can be gained or lost when water is being pumped above or below the level of the pump. NFPA 1001: 5.3 NFPA 6.3. *FFHB, 2E:* Page 291.

Question #12. As matter interacts, substances are formed, changed, and destroyed. Matter doesn't disappear; it merely changes form. NFPA 1001: 5.3; 6.3. *FFHB, 2E:* Page 79.

Question #13. Fire department standard operating procedures cover dynamic topics and must be reviewed often, at least once every three years. NFPA 1001: 5.1.1.1. *FFHB, 2E:* Page 35.

Question #14. 3. For instance, get water on exposed tanks to reduce internal pressure buildup and prevent the possibility of a BLEVE. 4c. For some fires, consider letting the fire burn itself out. NFPA 1001: 5.3.10. *FFHB, 2E:* Page 626.

Question #15. Advancing hoselines using a standpipe system involves two different hoseline evolutions. The first is the engine driver connecting to the fire department connection on the structure. The second is the hose crew connecting to the standpipe outlet and advancing the hoseline to attack the fire. NFPA 1001: 5.3. *FFHB, 2E:* Page 255.

Question #16. In newer systems, the protective systems are integrated into a single "smart" system that includes an annunciator panel, fire alarm control panel, and system override controls. NFPA 1001: 6.5.1. *FFHB, 2E:* Page 314.

Question #17. In fact, this deceptiveness has caused a great number of firefighter deaths and injuries over the years. Numbers of firefighters have been overrun in grassy fuels thinking they could "outrun" the fire when things went bad. NFPA 1001: 5.3.19. *FFHB, 2E:* Page 597.

Question #18. While inspecting cooking surfaces and hood and duct systems, firefighters should ensure that the extinguishing agent cylinders are charged and armed and the manual activation control is accessible. NFPA 1001: 5.5.1. *FFHB, 2E:* Page 668.

Question #19. Signs of a potential backdraft: puffing of smoke at seams and cracks of windows and doors NFPA 1001: 5.3.10;5.3.11. *FFHB, 2E:* Page 557.

Question #20. Procedures can be classified as formal or informal. Examples of informal procedures include a company officer's instructions at the beginning of a shift or on-the-job training a new member receives. NFPA 1001: 6.1.1.1. *FFHB, 2E:* Page 111.

Question #21. Dressing a knot is the practice of making sure all parts of the knot are lying in the properorientation to the other parts and look exactly as the pictures indicated. NFPA 1001: 5.1.1.1. *FFHB, 2E:* Page 426.

Question #22. When a sprinkler head is fused by heat, air is first discharged. As the air pressure drops below the pressure of the supply water, the clapper valve is opened and locked NFPA 1001: 6.5.1. *FFHB, 2E:* Page 319.

Question #23. The responsibility for firefighter safety is shared by three areas: the department itself (administration), the working teams, and the individual firefighter. NFPA 1001: 6.1.1.1. *FFHB, 2E:* Page 117.

Question #24. Salvage and overhaul tools and equipment must be cleaned and inspected like any other piece of firefighting equipment to make sure they are ready for the next emergency. NFPA 1001: 5.3.14; 5.3.13; 5.5.3. *FFHB, 2E:* Page 636.

Question #25. By-products of combustion can include water vapor, carbon particles, carbon monoxide, sulfur dioxide, and hydrogen cyanide. NFPA 1001: 5.3; 6.3. *FFHB, 2E:* Page 83.

Question #26. Hanging to dry.This is one of the eaiser methods of drying ropes. NFPA 1001: 5.3. *FFHB, 2E:* Page 443.

Question #27. Nozzle reaction is the force of nature that makes the nozzle move in the opposite direction of water flow. NFPA 1001: 5.3 NFPA 6.3. *FFHB, 2E:* Page 282.

Question #28. A tool that is too heavy cannot be moved fast enough to develop proper force. NFPA 1001: 5.3.4; 6.3.2. *FFHB, 2E:* Page 513.

Question #29. The follow-through figure-eight knot is very useful when attaching a utility or life safety line rope to an object that does not have a free end available. NFPA 1001: 5.1.1.1. *FFHB, 2E:* Page 433.

Question #30. Firefighters are responding to a fire in a structure that is Type V construction. The most common occupancy of this type of construction is residential. NFPA 1001: 5.3.10; 6.3.2. *FFHB, 2E:* Page 360.

Question #31. A reverse lay is the opposite of the forward lay with the supply line being dropped off at the fire location and the engine laying the hose toward the water source. NFPA 1001: 5.3.10. *FFHB, 2E:* Page 270.

Question #32. Wood is an organic material. Iron, granite, and quartz are inorganic materials. NFPA 1001: 5.3; 6.3. *FFHB, 2E:* Page 80.

Question #33. The shoulder toss is done by one firefighter and is used for covering large, unbreakable items. NFPA 1001: 5.3.14. *FFHB, 2E:* Page 643-644.

Question #34. Responders should understand basic recognition and identification skills. NFPA 472: 5.2.2 - 5.2.4. *FFHB, 2E:* Page 765.

Question #35. This type of unit must be equipped with an SCBA escape unit with duration of approximately five to ten minutes. NFPA 1001: 5.3.1. *FFHB, 2E:* Page 158.

Question #36. Cardiovascular conditioning activities are intended to raise the heart and breathing rates in order to strengthen the heart/lung relationship and reduce the chance of overexertion on the scene of an incident. NFPA 1001: 5.3; 6.3. *FFHB, 2E:* Page 727.

Question #37. Specific chemical information is important for a safe and informed response. NFPA472: 4.2.1 - 4.2.2. *FFHB, 2E:* Page 807.

Question #38. The safety triad includes personnel, equipment, and procedures. NFPA 1001: 6.1.1.1. *FFHB, 2E:* Page 111.

Question #39. NFPA 1901, Standard for Automotive Fire Apparatus, states that a pumper should have a permanently mounted fire pump with a capacity of at least 750 gallons per minute and that the pumper should carry no less than 300 gallons of water. NFPA 1001: 5.3.15. *FFHB, 2E:* Page 29.

Question #40. But be aware there may be additional locking devices out of view that may not be easily unlocked. NFPA 1001: 5.3.4; 6.3.2. *FFHB, 2E:* Page 532.

Question #41. Vandals can also put debris into the system, damaging the hydrant or the pump; the firefighter should check the outlet prior to connecting the hose and flush it if necessary. NFPA 1001: 5.3.15. *FFHB, 2E:* Page 216.

Question #42. Voids spaces within a collapsed area that are open and may be an area where someone could survive a building collapse. NFPA 1001: 6.4.2. *FFHB, 2E:* Page 506.

Question #43. Extinguishers in buildings should be checked every 30 days, and extinguishers on apparatus should be inspected each time the vehicle is inspected. NFPA 1001: 5.3.16. *FFHB, 2E:* Page 198.

Question #44. The maximum quantity of flammable and combustible liquids permitted to be stored inside a building varies according to the occupancy. NFPA 1001: 5.5.1. *FFHB, 2E:* Page 669.

Question #45. SRS devices that have not deployed should be considered "live." NFPA 1001: 6.4.1. *FFHB, 2E:* Page 488.

Question #46. Understanding protective clothing and its relationship to toxicology and health effects is important. NFPA 472: 5.3.3; 5.4.3. *FFHB, 2E:* Page 848.

Question #47. The testing of hose begins with a visual inspection of the hose coupling. This inspection should be done with the annual test and during routine reloading and reconnection of hose sections. As the hose is loaded, it should also be visually inspected for any type of damage. NFPA 1001: 6.5.3. *FFHB, 2E:* Page 274.

Question #48. Now it is easier to understand why operating a nozzle in the sitting or lying position is less stressful than the standing position; the ground helps absorb this reaction. NFPA 1001: 5.3 NFPA 6.3. *FFHB, 2E:* Page 292.

Question #49. She should use the command "raise" as this means to move a ladder from a horizontal position to a vertical angle. NFPA 1001: 5.3.6; 5.3.9; 5.3.10; 5.3.11; 5.3.12. *FFHB, 2E:* Page 375.

Question #50. There are many key historical figures that were involved in the development of the American fire service, including Benjamin Franklin, George Washington, Thomas Jefferson, John Hancock, and Samuel Adams. NFPA 1001: N/A. *FFHB, 2E:* Page 9.

Question #51. Is a fog stream the correct choice? If so, which width? Figure 19-19 (FFHB, 2E) Firefighters use a fog pattern to shield themselves from radiant heat. NFPA 1001: 5.3.10. *FFHB, 2E:* Page 604.

Question #52. Making equipment safe is addressed in three ways: equipment selection, equipment inspection and maintenance, and equipment application. NFPA 1001: 5.5.3; 5.5.4; 6.1.1.1; 6.5.2. *FFHB, 2E:* Page 112.

Question #53. For example, opening a bulkhead door will be quick and effective. It should be performed before a roof cut is initiated. NFPA 1001: 5.3.11. *FFHB, 2E:* Page 576.

Question #54. As a rule of thumb, only organic materials will burn. NFPA 1001: 5.3; 6.3. *FFHB, 2E:* Page 80.

Question #55. First responders should have a basic awareness of the threat of terrorism and basic response actions. NFPA 472: 4.2.1. *FFHB, 2E:* Page 917.

Question #56. Understanding protective clothing and its relationship to toxicology and health effects is important. NFPA 472: 5.3.3; 5.4.3. *FFHB, 2E:* Page 843.

Question #57. These valves must have a chain lock on them to prevent tampering. A wrench or a wheel controls these valves; a padlock and chain are used to lock them open. NFPA 1001: 6.5.1. *FFHB, 2E:* Pages 324 and 325.

Question #58. Understanding the regulations covering hazardous materials is important. NFPA 472: 4.2.1-4.2.3. *FFHB, 2E:* Page 748.

Question #59. It then naturally follows that the steeper the hill, the faster the fire will travel upward and the slower it will travel downward. NFPA 1001: 5.3.19. *FFHB, 2E:* Page 599.

Question #60. The biggest problem with PASS devices results when wearers simply forget to turn their units on. This simple mental lapse has contributed to numerous firefighter fatalities. NFPA 1001: 5.1.1.2. *FFHB, 2E:* Page 135.

Question #61. Atmospheres with oxygen concentrations below 19.5 percent are classified as oxygen-deficient atmospheres. NFPA 1001: 5.3.1. *FFHB, 2E:* Page 145.

Question #62. For instance, a 2-A extinguisher will put out twice the fire of a 1-A. NFPA 1001: 5.3.16. *FFHB, 2E:* Page 194.

Question #63. Fire load is measured in Btus and is used mainly in fire engineering and not in the construction industry. NFPA 1001: 5.3.10; 6.3.2. *FFHB, 2E:* Page 342-344.

Question #64. Firefighters arrive on the scene to find an adult male patient who appears to be unconscious. The scene is safe and the firefighters are using BSI. They begin to assess the patient. This can all be done because of implied consent. Patients who are unconscious or mentally impaired are provided emergency medical care under the assumption that they would want the care if they were conscious or mentally capable. NFPA 1001: 4.3. *FFHB, 2E:* Page 694.

Question #65. Responders should understand basic recognition and identification skills, and understand the dangers of explosives. NFPA 472: 5.2.2 - 5.2.4. *FFHB, 2E:* Page 761.

Question #66. Responders should understand the protective actions that are available to utilize. NFPA 472: 5.3.2 - 5.3.4. *FFHB, 2E:* Page 898.

Question #67. An overhand knot is generally used to secure the loose end of the working end after tying a knot. NFPA 1001: 5.1.1.1. *FFHB, 2E:* Page 427.

Question #68. Understanding of basic chemical and physical properties is important for the health and safety of emergency responders. NFPA 472: 5.2.2 - 5.2.4. *FFHB, 2E:* Page 800.

Question #69. Balloon frame construction allows fire to spread to an attic more rapidly. Fire can enter the wall space and run straight to the attic because the wood studs run from the foundation to the roof and floors are "hung" on the studs. NFPA 1001: 5.3.10; 6.3.2. *FFHB, 2E:* Page 356-357.

Question #70. Caution: In fact, in many cases, water could create more problems than the firefighting team cares to imagine. No one agent applies to all fires in this category, and, in some cases, it may not be the best solution to extinguish the fire at all. It may be better to let it burn itself out as in the case of some pesticides and gases. NFPA 1001: 6.3.1. *FFHB, 2E:* Page 626.

Question #71. Responders should have a basic understanding of the SARA regulation. NFPA 472: 4.2.1-4.2.3. *FFHB, 2E:* Page 745.

Question #72. All of these definitions have been used to describe combustion. NFPA 1001: 5.3; 6.3. *FFHB, 2E:* Page 76.

Question #73. Indirect fire attack is used to attack interior fires by applying a fog stream into a closed room or compartment, converting the water into steam to extinguish the fire. Firefighters apply the water at the doorway and then close the door, allowing the steam to put the fire out. NFPA 1001: 5.3 NFPA 6.3. *FFHB, 2E:* Page 289.

Question #74. Head pressure measures the pressure at the bottom of a column of water in feet. Head pressure can be gained or lost when water is being pumped above or below the level of the pump. This is also called elevation pressure and is usually rounded to 5 psi per floor when pumping water up or down in a building. NFPA 1001: 5.3 NFPA 6.3. *FFHB, 2E:* Page 291.

Question #75. While it may seem easy, stopping water flow from a sprinkler head requires practice to effectively stop the flow and establish a leak-free seal. The firefighter assigned to stop sprinkler flow at the head will get seriously wet. NFPA 1001: 6.5.1. *FFHB, 2E:* Page 326.

Question #76. Understanding various protective actions is important for responder health and safety. NFPA 472:5.4.1. *FFHB, 2E:* Page 861.

Question #77. Understanding various protective actions is important for responder health and safety. NFPA 472: 5.2.3 -5.2.4. *FFHB, 2E:* Page 869.

Question #78. Responders should understand the protective actions that are available to utilize. NFPA 472: 6.2.1 - 6.2.2. *FFHB, 2E:* Page 905.

Question #79. Firefighter Johnson should not pack all the PPE into its carrying bag at the beginning of his shift. All PPE should be present and positioned so that it may be rapidly donned. NFPA 1001: 5.3; 6.3. *FFHB, 2E:* Page 724.

Question #80. A distinct advantage of the PPV blower is ease of setup. If properly applied, the blower can do the job of several firefighters attempting to open multiple holes with ladders at different locations. NFPA 1001: 5.3.11. *FFHB, 2E:* Page 565.

Question #81. As the use of security gates and overhead doors increases, the power saw has become the tool of choice to remove the door or gate. NFPA 1001: 5.3.4; 6.3.2. *FFHB, 2E:* Page 520.

Question #82. He should use the command "extend" as this means to increase the length and reach of the ladder with the fly sections. NFPA 1001: 5.3.6; 5.3.9; 5.3.10; 5.3.11; 5.3.12. *FFHB, 2E:* Page 375.

Question #83. Understanding protective clothing and its relationship to toxicology and health effects is important. NFPA 472: 5.3.3; 5.4.3. *FFHB, 2E:* Page 843.

Question #84. The NFPA states that "95 percent of alarms shall be answered within 30 seconds, and in no case shall the initial operator's response to an alarm exceed 60 seconds." NFPA 1001: 5.2.1. *FFHB, 2E:* Page 49.

Question #85. The firefighter can expect panic hardware to be on the exits. Panic hardware is required on exits in all assembly, educational, and institutional occupancies. NFPA 1001: 5.5.1. *FFHB, 2E:* Page 664.

Question #86. The next firefighter places the hoseline over the left shoulder at the next coupling and begins to climb the ladder. NFPA 1001: 5.3. *FFHB, 2E:* Page 259.

Question #87. Specific chemical information is important for a safe and informed response. NFPA472: 4.2.1 - 4.2.2. *FFHB, 2E:* Page 817.

Question #88. Firefighters are fighting a fire in a building that was constructed with composite trusses made of both wood and steel. This is dangerous for firefighters as the mixture of steel and wood can cause rapid collapse because steel expands faster than wood, causing stress at the intersections of the two materials. NFPA 1001: 5.3.10; 6.3.2. *FFHB, 2E:* Page 350-351.

Question #89. Drafting water from a lake or the sea is accomplished by taking advantage of this atmospheric pressure. NFPA 1001: 5.3.15. *FFHB, 2E:* Page 213.

Question #90. The wildland PPE ensemble is designed to be worn over undergarments. These undergarments (long-sleeve t-shirt, pants, and socks) should be 100 percent cotton or of a fire-resistive material. NFPA 1001: 5.1.1.2. *FFHB, 2E:* Page 130.

Question #91. There are additional legal considerations when responding to hazardous materials releases. NFPA 472: 4.2.1. *FFHB, 2E:* Page 749.

Question #92. In the two-story home, the bedrooms will generally be on the upper floor (not always) with possibly one of them downstairs. NFPA 1001: 5.1.1.1. *FFHB, 2E:* Page 612.

Question #93. Understanding various protective actions is important for responder health and safety. NFPA 472: 5.2.4. *FFHB, 2E:* Page 853.

Question #94. The basic goal of salvage operations is to remove the harmful atmosphere from the material or protect the material from the harmful atmosphere. NFPA 1001: 5.3.14. *FFHB, 2E:* Page 633-634.

Question #95. Specific chemical information is important for a safe and informed response. NFPA 472: 4.2.1 - 4.2.2. *FFHB, 2E:* Page 808.

Question #96. All these methods can be used to control external bleeding. NFPA 1001: 4.3. *FFHB, 2E:* Page 711-712.

Question #97. Residential sprinklers are designed for life safety and not necessarily to protect property. NFPA 1001: 6.5.1. *FFHB, 2E:* Page 314.

Question #98. The engine has to be able to position itself close enough to place the hard sleeve and strainer into the water or else reach the dry hydrant location. NFPA 1001: 5.3. *FFHB, 2E:* Page 221.

Question #99. When AFFF was introduced, one of its additional advantages was that it did not require a special foam nozzle for application. NFPA 1001: 5.3 NFPA 6.3. *FFHB, 2E:* Page 302.

Question #100. Telecommunicators must be able to prioritize incoming lines in order to ensure that the most important call gets the fastest attention. Incoming telephone calls should be answered in the following priority: 1. emergency or 9-1-1 lines 2. direct lines 3. business or administrative lines NFPA 1001: 5.2.1; 5.2.2. *FFHB, 2E:* Page 51.

Phase Two, Exam Two

1. The physical state of matter (solid, liquid, or gas) can affect combustion.
   a. True
   b. False

2. Class K and Class B fires are similar.
   a. True
   b. False

3. According to the NFPA, two of the HazMat-related tasks that are considered defensive are containment and confinement.
   a. True
   b. False

4. A material that is heavy and dense will be a less effective thermal conductor than a material that is light and less dense.
   a. True
   b. False

5. Any undeployed SRS device should be considered potentially "live" until disconnected.
   a. True
   b. False

6. A metal oxide sensor in a home carbon monoxide detector has the ability to detect propane in addition to CO.
   a. True
   b. False

7. You are doing morning equipment checks when you find the rotary saw, disks, and fuel have been moved to a single compartment. You should immediately move the fuel to another compartment.
   a. True
   b. False

8. You are fighting an interior room and contents fire. You find that your face piece is cracked. You know that positive pressure inside the mask should protect you as you leave the structure.
   a. True
   b. False

9. Protein foam is made from animal blood and salt.
   a. True
   b. False

10. The most common cause of harm for solids is inhalation.

   a. True

   b. False

11. When you are entering a high-rise building fire, you should always opt to use the stairs as elevators are unsafe in fire conditions for any reason.

   a. True

   b. False

12. Flash point is extremely important when considering flammable liquid extinguishment.

   a. True

   b. False

13. It is now possible to preplan IDLH atmospheres.

   a. True

   b. False

14. Mortar relies on compressive forces to give a masonry wall strength. Mortar mixes have little to no tensile or shear strength.

   a. True

   b. False

15. Because regular training is the single most important step in firefighter safety, the firefighter must strive to retain the information and skills that are presented in training sessions.

   a. True

   b. False

16. You are fighting an interior room and contents fire when your partner's low air warning goes off. In this situation, you should send the firefighter out and ask him to tell one of the "two out" firefighters to come in and take over his duties.

   a. True

   b. False

17. It is more effective for firefighters to determine the type of construction and fire-resistive rating once a building is complete as opposed to viewing the building during construction.

   a. True

   b. False

18. If firefighters use a manila rope for water rescue, they can expect the rope strength to be permanently reduced.

   a. True

   b. False

19. You are on vacation in Florida, and see a residential structure fire operation across the street from your hotel. You notice the hydrant looks different from the ones that you have seen in Colorado. The difference is that areas like Florida can have dry barrel hydrants, while areas such as Colorado would not be able to use them.

    a. True

    b. False

20. For a fire involving gasoline, a Class A fire extinguisher is acceptable.

    a. True

    b. False

21. You arrive on scene of a residential structure fire and see smoke coming out of the windows. You know that the smoke is most likely meeting resistance while moving vertically.

    a. True

    b. False

22. The failure of one structural element will typically cause the loads to be transferred to other elements.

    a. True

    b. False

23. While searching a large commercial department store, you decide to forego a guideline in favor of doing a left hand search. This is a safe and accepted practice that will save valuable time.

    a. True

    b. False

24. The term corrosive only describes acids.

    a. True

    b. False

25. Failure to use PPE or to use it improperly may result in injury or fatal effects.

    a. True

    b. False

26. You are on scene of a structure fire and find it necessary to "make the window a door." This is acceptable, if necessary, but should be avoided otherwise.

    a. True

    b. False

27. You are fighting a dumpster fire and your SCBA regulator fails. The fire was large, but you have knocked it back and almost have it extinguished. Your action should be to remove your regulator to finish the job and then let another firefighter do the overhaul.

    a. True

    b. False

28. Two firefighters are conducting salvage operations in a room within a structure. They should select the largest or most central item in the room and move all smaller items to that area before deploying salvage covers to protect the items.
    a. True
    b. False

29. The predominant hazard to firefighters when wearing any type of PPE is
    _____.
    a. visibility
    b. headache
    c. heat stress
    d. acute exposure

30. After donning PPE, you should use the _____ to ensure complete and appropriate coverage.
    a. mirror
    b. team check
    c. memory method
    d. repetitive line of check

31. You have been tasked with search and rescue on a residential structure fire. It is 3 A.M. The home is a two-story home. It can generally be assumed that your greatest rescue potential lies _____.
    a. on the uppermost floor
    b. in the basement
    c. on ground level
    d. both a and b

32. You are a venting a flat-roofed structure. After leaving the ladder, you should
    _____.
    a. cut the vent hole
    b. notify command you are on the roof
    c. make a rectangular inspection cut
    d. make a triangular inspection cut

33. Your engine driver/operator tells you she has been given the task of charging the deck gun. You know that _____.
    a. two or more firefighters will be needed to advance this line
    b. she wants you out of the way
    c. massive water is needed on this operation
    d. she is setting up for an offensive attack

34. Nozzle reaction is _____.
    a. the movement of the nozzle opposite from the direction of the water flow
    b. the movement of the nozzle in the same direction as the water flow
    c. dangerous to persons weighing less than the nozzle operator
    d. inversely proportionate to the gpm

35. When transmitting messages on mobile and portable radios, firefighters should hold the microphone _____ inches from the mouth and at a _____ -degree angle for a clear transmission.

    a. 2-3; 45

    b. 1-2; 90

    c. 1-2; 45

    d. 2-3; 90

36. Firefighter readiness for an incident includes which of the following?

    a. Wearing PPE and SCBA as appropriate

    b. Mental fitness

    c. Physical fitness

    d. All of the above

37. A rehabilitation area at the fireground is a _____ effort to prevent injury.

    a. reactive

    b. proactive

    c. both a and b

    d. none of the above

38. You and your partner are conducting a search operation in a smoke-filled residential structure. You find two unconscious men in the basement near a pool table. The men are both of equal size and appear to weigh in excess of 200 lbs. Your partner weighs 118 lbs and does not have a lot of upper body strength. You suggest she use a _____ to move the second man to the bottom of the stairs.

    a. foot drag

    b. a webbing sling drag

    c. firefighter's carry

    d. seat carry

39. While you are on a firefighting operation, you become thirsty and there is no bottled water available to you. You consider your knowledge of water systems and decide you _____.

    a. can drink water from a city fire hydrant, no matter the system pressure

    b. cannot drink water from a city fire hydrant as it is untreated

    c. can drink water from a city fire hydrant, as long as the system has a residual pressure of 20 psi

    d. can drink water from a city fire hydrant, as long as the system has a static pressure or 20 psi

40. You find it necessary to tie a life line around a large metal light pole. The knot suitable for this operation would be a _____.

    a. basic figure-eight knot

    b. follow-through figure-eight knot

    c. bowline knot

    d. follow-through bowline knot

41. You are a firefighter who has been on a month-long vacation. Upon return to work, you realize you can't recall when your work uniform was last cleaned. You recall that NFPA 1581 requires a uniform be cleaned _____ as a minimum, so you make sure you wash all pieces of your work uniform immediately.

    a. every month

    b. every six months

    c. every week

    d. yearly

42. You are fighting a fire in an industrial area of town at a business that produces pesticides. One of the tanks is burning freely. Your IC is gathering information and resources about the product. As a firefighter, you know that your strategy might be _____.

    a. an indirect attack

    b. a direct attack

    c. a transitional attack

    d. to let it burn

43. You are doing area familiarization in a rural area and notice there are no fire hydrants. It would be prudent for you to ascertain if there are any of these available for your use.

    a. Tanks

    b. Ponds

    c. Cisterns

    d. All of the above

44. You are assisting another crew with cleanup and are helping them get their apparatus back in service. The life-safety ropes have been washed and are still wet. Probably the best and one of the easiest methods of drying the ropes is _____.

    a. laying them flat

    b. machine drying them

    c. hanging them

    d. Synthetic ropes require no drying time. They can be bagged and put back on the apparatus.

45. A large warehouse fire has a total of 22 firefighters working in the Operations Section. Firefighter Smith says that it is acceptable to split the 22 firefighters into three divisions for an effective span of control. Firefighter Taylor says that the 22 firefighters need to be split into at least four divisions for an effective span of control. Who is correct?

    a. Firefighter Smith is correct.

    b. Firefighter Taylor is correct.

    c. Both are correct.

    d. Neither are correct.

46. Two firefighters are raising an extension ladder to a structure. This ladder will be used by many firefighters to access the structure. Where can the firefighters find the maximum load capacity of the ladder in order not to overload it?

    a. On the manufacturer's label affixed to the ladder

    b. Engraved in the bottom rung of the ladder

    c. Engraved in the top rung of the ladder

    d. It is not located on the ladder. The firefighters have to remember the information.

47. You are fighting a residential structure fire and command has just declared the fire knocked-down. Unless otherwise directed, you should immediately prepare for _____ operations.

    a. salvage

    b. attack

    c. overhaul

    d. rehab

48. Which of the following methods of opening an airway would be used on a patient who firefighters suspect has been the victim of a traumatic injury?

    a. The head-tilt, chin-lift method

    b. The jaw thrust method

    c. Any of these methods

    d. None of the above

49. Which tool would a firefighter select to help remove debris from a structure during overhaul operations.

    a. Catch-all

    b. Water vacuum

    c. Floor runner

    d. Carry-all

50. To protect a firefighter's hair, and to offer protection for the straps of a SCBA mask, a firefighter should don the _____.

    a. helmet

    b. protective hood

    c. SCBA

    d. coat

51. Heat comes from four basic sources. Friction would be an example of what type of heat source?

    a. Chemical

    b. Mechanical

    c. Electrical

    d. Nuclear

52. After a ladder was used at a structure fire incident, what can a firefighter look for to see if the ladder has been exposed to a potentially damaging heat level?

    a. The guides/channels

    b. The halyard

    c. The sensor label

    d. The tie-rods

53. Which of the following is a commonly used tool in salvage operations?

    a. Pike pole

    b. Water vacuum.

    c. Rubbish hook

    d. Axe

54. Which of the following American leaders played an important role in the development of the American fire service?

    a. Benjamin Franklin

    b. George Washington

    c. Thomas Jefferson

    d. All of the above

55. A simple way to teach the general public the different types of extinguishers is to _____.

    a. have them look for the symbols on the extinguisher

    b. have them learn the different colors of the extinguishers

    c. learn the shape of the extinguishers

    d. there is no simple way

56. A SARA Title III facility should be able to provide a _____ to emergency responders.

    a. MSDS

    b. DOT ERG

    c. NIOSH pocket guide

    d. Offloading truck

57. You are assigned to "wrap the hydrant" on the way into the fire. The engine stops, you pull off the layout load from the hose bed, and wrap the hydrant. Training has taught you that your next step is to _____.

    a. connect the hose

    b. position yourself

    c. notify the IC

    d. tell the engine to move ahead

58. Which of the following is not true concerning school evacuation drills?

 a. Specific exits should be assigned to groups of classrooms and all exits should be used to provide for even distribution of exiting occupants.

 b. Emergency evacuation plans should be drawn in graphic form and posted at each exit.

 c. The focus of the drill should be disciplined control and order. Speed will result from properly planned and supervised evacuations.

 d. The fire department should be present whenever possible at evacuation drills.

59. Firefighter Carey notices that her co-worker, Firefighter Nally, has become increasingly withdrawn, inattentive, and depressed since returning to the station after a fire. Firefighter Nally might be experiencing _____.

 a. critical incident stress

 b. heat stress

 c. overexertion

 d. malnutrition

60. Which of the following is a commonly used tool in overhaul operations?

 a. Pike pole

 b. Salvage cover

 c. Catch-all

 d. Hammer

61. When responding to a fire alarm in a city business, firefighters should don a _____ ensemble.

 a. swift-water

 b. structural

 c. wildland

 d. proximity

62. How important are the first few minutes at a hazardous materials incident?

 a. Not important

 b. Only for rail cars

 c. Only for long-term incidents

 d. They guide the rest of the response.

63. When an extension ladder exceeds _____ feet, it is required to be equipped with staypoles.

 a. 20

 b. 30

 c. 40

 d. 50

64. What is the highest priority when responding to a fire involving explosives?
    a. Life safety
    b. Adequate hoselines
    c. Spill control
    d. Temperature control

65. You find yourself needing to cut a hole in a wooden floor for an emergency escape. You estimate you have about five minutes to leave the area you are in and you have only an axe with which to work. Your first step is to _____.
    a. sound the floor for the joists
    b. cut through the subfloor
    c. request a chainsaw from command
    d. use your PASS

66. You are forcing entry into a garage. You have managed to open the overhead door. To ensure your safety, the next tool you use should be _____.
    a. a rotary saw
    b. locking pliers
    c. a three-foot pike pole
    d. none of the above

67. Many units from the Delmar Fire Department are on the scene of a fire in an apartment building. The incident commander wants to check on the availability of getting additional hose from the fire department storage area to use on the scene. Who should he contact?
    a. The operations section chief
    b. The planning section chief
    c. The logistics section chief
    d. The finance/administration section chief

68. You are attending church when a water motor gong sounds. As you are evacuating the building, you think and don't recall ever seeing an annunciator panel. Your reaction should be to _____.
    a. warn people to stay clear of the main exit as that is where firefighters will enter the structure
    b. phone in the alarm to dispatch once you have evacuated as you suspect it is a local alarm only
    c. grab a fire extinguisher and sweep the building looking for the fire
    d. gather all the children and hold them in one place for interrogation and lecture about setting off false alarms

69. When setting up a tender shuttle operation, you direct that the _____ be set up first.
    a. fill site
    b. dump site
    c. flow site
    d. drafting area

70. In a wildland operation where you have driven the fire back, you find you have run out of available hose and can no longer reach the fire. Your preferred choice of action is _____.

    a. have the driver/operator advance the engine

    b. quickly extend the hoseline

    c. have another crew pull an additional line

    d. turn the operation over to another apparatus

71. Which category of WMD agents can only be treated by moving the victims to fresh air?

    a. Blister agents

    b. Nerve agents

    c. Nuclear

    d. Irritants

72. To determine how many additional lines can be connected to a fireground operation, it is vital that you remember to check the _____ pressure before any lines are opened.

    a. static

    b. residual

    c. flow

    d. nozzle

73. Your inspection cut reveals a very thick roof. Your first operational duty in this ventilation should be to _____.

    a. make an inspection cut

    b. make a louvered cut

    c. remove the roof covering

    d. make an expandable cut

74. You are holding a patient's head after extricating her from a motor vehicle accident. The patient complained of severe neck pain. Your partners lay a backboard out next to the patient.One rescuer assumes the position at the patient's torso, another at the patient's hips, and another at the patient's feet._____ will direct the movement of the patient onto the backboard.

    a. The rescuer at the torso

    b. You

    c. The rescuer at the feet

    d. The rescuer at the hips

75. You are preparing to hoist a pike pole to a second story roof. You are on the ground tying the rope. From your training, you remember that the two ways of tying a clove hitch are _____ or around an object.

    a. in the open

    b. in the air

    c. bowline

    d. Becket bend

76. Firefighter Nelson is helping emergency medical service personnel set up a rehabilitation area on the scene of a structure fire. Which of the following is a key element that Firefighter Nelson should include in the rehabilitation area?
    a. Active cooling
    b. Hydration
    c. Nourishment
    d. All of the above

77. You are called to a residence to investigate a malfunctioning smoke detector. Upon arrival on scene, the homeowner tells you she keeps hearing the smoke detector "beep." You should _____.
    a. investigate and offer to change the smoke detector batteries
    b. investigate, but be prepared to blow the dust out of the smoke detector
    c. don your airpack and investigate with a gas detector
    d. offer to install a new smoke detector

78. A firefighter conducting a fire safety inspection finds a business that stores flammable and combustible liquids inside the facility. He knows that quantities less than _____ gallons must be stored in approved safety containers and containers with a capacity greater than _____ gallons may not be stacked.
    a. 3; 30
    b. 5; 30
    c. 3; 50
    d. 5; 50

79. Methane has a vapor density of 0.6. If Methane escapes its container, it can be expected to do what?
    a. Rise into the air
    b. Stay level with the container
    c. Collect in low-lying areas
    d. None of the above

80. You are fighting an interior room and contents fire. Visibility is very limited; however you notice a window in the room. You can employ _____ ventilation to clear the smoke and continue your attack of the fire.
    a. water
    b. hydraulic
    c. natural
    d. horizontal

81. The four elements of a fire stream are the _____.
    a. pump, water, hose, and nozzle
    b. source, hose, pressure, and nozzle
    c. psi, gpm, length of hose, and friction loss
    d. type of source, diameter of hose, length of hose, and gallons per minute

82. You answer a shift recall to your station and realize you left your structural gear at home.Luckily, you are stationed at the airport station and have your proximity gear. A fire alarm call drops and you don your gear and get on the engine. It is important for you to remember that your proximity gear is designed for _____.

   a. high-heat and insulation standards

   b. only aircraft fires

   c. only Class B fires

   d. total fire envelopment

83. It is a hot afternoon and the temperature is nearing 95°F. You and your engine company are responding to a reported MVA—car vs. semi with occupants trapped. In this case, which attitude should you adopt?

   a. Wait-and-see

   b. Double-dress

   c. Dress-down

   d. Dress-up

84. What protective device greatly adds to a firefighter's chance for survival when confronted with potential chemical exposure?

   a. Boots

   b. Gloves

   c. APR

   d. SCBA

85. You are preparing to hoist an axe to the roof of a structure. Before you signal that the axe is ready to be hoisted, you tie a(n) _____ in the working end.

   a. overhand knot

   b. clove hitch

   c. Becket bend

   d. double Becket bend

86. When fighting a wildland fire, it is generally much safer to be _____ the fire.

   a. upslope from

   b. downslope from

   c. parallel to

   d. perpendicular to

87. The _____ nozzle was designed for aircraft operations.

   a. fog

   b. piercing

   c. ARFF ram

   d. straight bore

88. You are the nozzle operator advancing a hoseline toward an automobile fire. The engine compartment is fully engulfed. You should approach the car from the _____.

a. rear

b. front

c. side

d. top

89. The information provided to an emergency communications center by mobile or portable radio must be _____.

a. accurate

b. clear

c. complete

d. all of the above

90. You are fighting an outdoor propane tank fire with your two partners. There is another hoseline advancing with three more firefighters and you are working in tandem. As you approach, you notice the heat appears to be increasing. You should _____.

a. retreat

b. decrease your fog pattern

c. increase your fog pattern

d. wait for the other team to advance

91. Many jurisdictions are working together to fight a wildfire on a mountain. The incident commander wishes to review the incident action plan. Who will she contact to get this information?

a. The operations section chief

b. The planning section chief

c. The logistics section chief

d. The finance/administration section chief

92. On the scene of a commercial structure fire, you are connecting hose to the Siamese fire department connection. You always connect to the left outlet first _____.

a. to allow better access for using the spanner wrench to tighten the coupling

b. as some outlets only have clappers on the right side

c. both of these

d. neither of these

93. You and two other firefighters are advancing a 2 1/2" hoseline into a structure that has flames and smoke showing. It is a difficult advance as there are many turns in a long hallway. You arrive at the fire and are backing up the firefighter at the nozzle. Your partner behind you taps your shoulder and points to his facemask. He turns to leave as he activates his PASS device and his bypass valve. You should _____.

    a. get your other partner's attention immediately

    b. relay to command that one member of your company is exiting the structure

    c. immediately turn and follow him out of the structure

    d. stop him, use your buddy breathing attachment, and continue backing up the nozzleman

94. You are filling an SCBA bottle from the Cascade unit in your station. You notice the bottle is heating rapidly while you are filling it. You should _____.

    a. stop filling the bottle immediately

    b. check the supply air to ensure it is not overheated

    c. reduce the airflow at the bottle

    d. reduce the fill rate

95. If a firefighter exhausts a portable fire extinguisher on a fire without extinguishing the fire, she should _____.

    a. grab another extinguisher, regardless of its rating or type, and continue to fight the fire

    b. grab another extinguisher of the same rating and continue to fight the fire

    c. search for a larger fire extinguisher to fight the fire

    d. use a hoseline or other methods suitable for a large fire

96. When you are working a fireline on a wildland fire, you wear leather boots that rise well above your ankle. This is for protection against _____.

    a. snake bites

    b. ankle injuries

    c. foot fatigue

    d. all of the above

97. Oxidizers and organic peroxides can have which type of characteristics?

    a. They create a lack of oxygen.

    b. They may be explosive.

    c. They are typically nonhazardous.

    d. They are never transported.

98. You are at a residence with light smoke showing. The front door has a window next to it. You break the window and unlock the door from the inside. However, you still cannot open the door. What is the most likely problem?

    a. Door swelling

    b. Additional locking devices

    c. Furniture

    d. Debris

99. You have responded to a call of a possible structure fire in a single-family residence. You arrive on scene and find smoke showing in a Spanish style home. The home has bars over the windows and security gates—all of which are locked. You decide your best option is to use a _____ for access.

a. Bam Bam

b. pick head axe

c. bolt cutters

d. rotary saw

100. While fighting a small fire with a portable extinguisher, you notice the fire is spreading with each sweep you make. A possible reason is you are _____.

a. using the wrong type of extinguisher

b. not aiming the extinguisher at the base of the fire

c. too close to the base of the fire

d. all of these

Phase II, Exam II: Answers to Questions

1. T	26. T	51. B	76. D
2. T	27. F	52. C	77. C
3. T	28. T	53. B	78. B
4. F	29. C	54. D	79. A
5. F	30. B	55. A	80. B
6. T	31. A	56. A	81. A
7. T	32. D	57. B	82. A
8. T	33. C	58. B	83. C
9. T	34. A	59. A	84. D
10. F	35. C	60. A	85. A
11. F	36. D	61. B	86. B
12. T	37. A	62. D	87. B
13. F	38. B	63. C	88. C
14. T	39. A	64. A	89. D
15. T	40. B	65. A	90. C
16. F	41. A	66. B	91. B
17. F	42. D	67. C	92. C
18. T	43. D	68. B	93. A
19. F	44. C	69. B	94. D
20. F	45. B	70. B	95. D
21. T	46. A	71. D	96. D
22. T	47. C	72. A	97. B
23. F	48. B	73. C	98. B
24. F	49. D	74. B	99. D
25. T	50. B	75. A	100. D

Phase II, Exam II:
Rationale & References for Questions

Question #1. The physical state of matter (solid, liquid, or gas) can affect combustion. NFPA 1001: 5.3; 6.3. *FFHB, 2E:* Page 98.

Question #2. Class K fuels are similar to Class B fuels but involve high-temperature cooking oils and, therefore, have special characteristics. NFPA 1001: 5.3.16. *FFHB, 2E:* Page 186.

Question #3. First responders should understand their role when involved in a chemical release event. NFPA 472: 5.3.2 - 5.3.4. *FFHB, 2E:* Page 892.

Question #4. A material that is heavy and dense will be a more effective thermal conductor than a material that is light and less dense. NFPA 1001: 5.3; 6.3. *FFHB, 2E:* Page 97.

Question #5. SRS devices that have not deployed should be considered "live." NFPA 1001: 6.4.1. *FFHB, 2E:* Page 488.

Question #6. Responders should understand the protective actions that are available to utilize. NFPA 472: 6.2.1 - 6.2.2. *FFHB, 2E:* Page 908.

Question #7. Streetsmart Tip Gasoline or other hydrocarbon fuels will break down the bonding material used in the manufacture of abrasive disc blades. NFPA 1001: 5.5.3. *FFHB, 2E:* Page 517.

Question #8. A constant supply of air is delivered, pressurizing the face piece, keeping toxic gases from entering. NFPA 1001: 5.3.1. *FFHB, 2E:* Page 144.

Question #9. Protein foam is made from chemically broken down natural protein materials, such as animal blood that have metallic salts added for foaming. NFPA 1001: 5.3 NFPA 6.3. *FFHB, 2E:* Page 296.

Question #10. Understanding of basic chemical and physical properties is important for the health and safety of emergency responders. NFPA 472: 5.2.2 - 5.2.4. *FFHB, 2E:* Page 795.

Question #11. If an elevator is utilized, the firefighters will attempt to stop and get off the elevator two or more floors below the reported fire floor and walk the stairs the rest of the way. NFPA 1001: 5.3.10. *FFHB, 2E:* Page 616.

Question #12. Flash point was mentioned, and that characteristic of every flammable liquid will make a very big difference in a material's storage, handling, and extinguishment. NFPA 1001: 5.3.7. *FFHB, 2E:* Page 601.

Question #13. Because it is impossible to predetermine all IDLH atmospheres, this policy must include operations during interior or exterior fire attack, such as structure, vehicle and dumpster fires, below-grade or confined space rescue, and hazardous materials incidents. NFPA 1001: 5.3.1. *FFHB, 2E:* Page 143.

Question #14. Mortar relies on compressive forces to give a masonry wall strength. Mortar mixes have little to no tensile or shear strength. NFPA 1001: 5.3.10; 6.3.2. *FFHB, 2E:* Page 350.

Question #15. Because regular training is the single most important step in firefighter safety, the firefighter must strive to retain the information and skills that are presented in training sessions. NFPA 1001: 6.1.1.1. *FFHB, 2E:* Page 114.

Question #16. Firefighters must always work in teams of two or more when entering an involved structure for any reason (i.e., interior firefighting, search and rescue, ventilation.) In addition to this search team, a minimum of two firefighters must be standing by immediately outside in full protective clothing and SCBA with a charged hoseline (although the charged hoseline is not required, it is a good logical practice) ready to come in and assist the search team should a problem develop. NFPA 1001: 5.3.9. *FFHB, 2E:* Page 462.

Question #17. It is more effective for firefighters to determine the type of construction and fire-resistive rating during a building's construction as opposed to viewing the building once it is complete. NFPA 1001: 5.3.10; 6.3.2. *FFHB, 2E:* Page 341.

Question #18. In actuality, this soaking reduced its strength by approximately 50% which it never regained. NFPA 1001: 5.1.1.1. *FFHB, 2E:* Page 420.

Question #19. Wet barrel hydrants have water in the barrel up to the valves of each outlet. They are used in areas that are not subject to freezing temperatures, primarily California and Florida. NFPA 1001: 5.3.15. *FFHB, 2E:* Page 206.

Question #20. Class B fires involve flammable and combustible liquids, gases, and greases. Common products are gasoline, oils, alcohol, propane, and cooking oils. NFPA 1001: 5.3.16. *FFHB, 2E:* Page 185.

Question #21. Note If a large volume of smoke is coming from the windows, it is a good indicator that the smoke and heat are meeting resistance in the attempt to move vertically. NFPA 1001: 5.3.11. *FFHB, 2E:* Page 552.

Question #22. The failure of one structural element will typically cause the loads to be transferred to other elements. NFPA 1001: 5.3.10; 6.3.2. *FFHB, 2E:* Page 347.

Question #23. However, this is not the case in larger mercantile, commercial, or industrial occupancies. NFPA 1001: 5.3.9. *FFHB, 2E:* Page 463.

Question #24. Understanding of basic chemical and physical properties is important for the health and safety of emergency responders. NFPA 472: 5.2.2 - 5.2.4. *FFHB, 2E:* Page 799.

Question #25. Understanding health and risk factors is important. NFPA 472: 5.3.3; 5.4.3. *FFHB, 2E:* Page 833.

Question #26. At the same time, firefighters must remember not to break glass unnecessarily and create a safety hazard. NFPA 1001: 5.3.4; 6.3.2. *FFHB, 2E:* Page 542.

Question #27. Trash can produce toxic gases and, in many cases, items discarded can be hazardous and sometimes even explosive. The unprepared firefighter can be injured or killed fighting these seemingly innocent fires. NFPA 1001: 5.3.10. *FFHB, 2E:* Page 626.

Question #28. Two firefighters are conducting salvage operations in a room within a structure. They should select the largest or most central item in the room and move all smaller items to that area before deploying salvage covers to protect the items. NFPA 1001: 5.3.14. *FFHB, 2E:* Page 642.

Question #29. Understanding protective clothing and its relationship to toxicology and health effects is important. NFPA 472: 5.3.3; 5.4.3. *FFHB, 2E:* Page 847.

Question #30. Practice "team checks."Firefighters should check each other's PPE for readiness. NFPA 1001: 5.1.1.2. *FFHB, 2E:* Page 138.

Question #31. In the two-story home, the bedrooms will generally be on the upper floor (not always) with possibly one of them downstairs. NFPA 1001: 5.1.1.1. *FFHB, 2E:* Page 612.

Question #32. The first operation that needs to be accomplished on a flat roof is the inspection cut. Then the triangular inspection cut is completed with another cut. NFPA 1001: 5.3.11. *FFHB, 2E:* Page 580.

Question #33. Master streams or heavy appliances are non-hand-held water applicators capable of flowing more than 350 gallons of water per minute. The wagon pipe is a permanently mounted master stream device on an engine that has either pre-piped water connection or needs a short section of hose to connect it to the pump. Some departments refer to wagon pipes as a "deck gun." NFPA 1001: 5.3. *FFHB, 2E:* Page 271.

Question #34. Nozzle reaction is the force of nature that makes the nozzle move in the opposite direction of the water flow. NFPA 1001: 5.3 NFPA 6.3. *FFHB, 2E:* Page 282.

Question #35. When transmitting messages on mobile and portable radios, firefighters should hold the microphone 1-2 inches from the mouth and at a 45-degree angle for a clear transmission. NFPA 1001: 5.2.3. *FFHB, 2E:* Page 65.

Question #36. A firefighter's readiness includes not only wearing PPE and SCBA as appropriate, but also being ready mentally and physically. NFPA 1001: 6.1.1.1. *FFHB, 2E:* Page 119.

Question #37. A rehabilitation area at the fireground is a reactive effort (intervention) to prevent injury. NFPA 1001: 6.1.1.1. *FFHB, 2E:* Page 110-111.

Question #38. The webbing drag enables a rescuer who is significantly smaller than the victim to perform a rescue. NFPA 1001: 5.3.9. *FFHB, 2E:* Page 472.

Question #39. After treatment for drinking, water goes into the distribution system or water mains. The mains are divided into feeders and distribution lines. Primary feeders supply secondary feeders and then distribution lines. Distributor lines are the water mains with the building connections and fire hydrants, and for fire protection; they range in size from 6 to 16 inches. NFPA 1001: 5.3.15. *FFHB, 2E:* Page 205.

Question #40. The follow-through figure-eight knot is very useful when attaching a utility or life safety line rope to an object that does not have a free end available. NFPA 1001: 5.1.1.1. *FFHB, 2E:* Page 433.

Question #41. NFPA 1581, Standard on Fire Department Infection Control Program, requires that clothing be cleaned every six months as a minimum. NFPA 1001: 5.1.1.2. *FFHB, 2E:* Page 136.

Question #42. Caution: In fact, in many cases, water could create more problems than the firefighting team cares to imagine. No one agent applies to all fires in this category, and, in some cases, it may not be the best solution to extinguish the fire at all. It may be better to let it burn itself out as in the case of some pesticides and gases. NFPA 1001: 6.3.1. *FFHB, 2E:* Page 626.

Question #43. An additional source of water can come from a developed source of water such as water tanks, ponds, and cisterns. NFPA 1001: 5.3.15. *FFHB, 2E:* Page 205.

Question #44. Hanging to dry. This is one of the easier methods of drying ropes. NFPA 1001: 5.3. *FFHB, 2E:* Page 443.

Question #45. Firefighter Taylor is correct. The range for effective span of control in the Incident Management System is 3 to 7 firefighters. If there are 22 total firefighters, they will need to be split into at least four divisions to properly accomplish an effective span of control. NFPA 1001: 6.1.1.1; 6.1.1.2. *FFHB, 2E:* Page 36.

Question #46. The maximum load for a ladder can be found on the manufacturer's label affixed to the ladder. NFPA 1001: 5.3.6; 5.3.9; 5.3.10; 5.3.11; 5.3.12. *FFHB, 2E:* Page 392.

Question #47. As soon as the fire is extinguished, the overhaul begins. NFPA 1001: 5.3.10. *FFHB, 2E:* Page 610.

Question #48. If the patient has experienced a traumatic injury, the method for opening the airway is the jaw thrust. NFPA 1001: 4.3. *FFHB, 2E:* Page 702.

Question #49. A carry-all is used to help remove debris from a structure during overhaul operations. NFPA 1001: 5.3.14. *FFHB, 2E:* Page 649.

Question #50. Hoods are made of fire-resistive, form-fitting cloth that protects the face, ears, hair, and neck in areas not covered by the helmet, earflaps, and coat collar. NFPA 1001: 5.1.1.2. *FFHB, 2E:* Page 129.

Question #51. Heat comes from four basic sources. Friction would be an example of a mechanical source of heat. NFPA 1001: 5.3; 6.3. *FFHB, 2E:* Page 83.

Question #52. After a ladder was used at a structure fire incident, a firefighter should look at the sensor label on a ladder to see if the ladder has been exposed to a potentially damaging heat level. NFPA 1001: 5.3.6; 5.3.9; 5.3.10; 5.3.11; 5.3.12; 5.5.3. *FFHB, 2E:* Page 372.

Question #53. A water vacuum is a commonly used tool in salvage operations. NFPA 1001: 5.3.14. *FFHB, 2E:* Page 635.

Question #54. There are many key historical figures that were involved in the development of the American fire service, including Benjamin Franklin, George Washington, Thomas Jefferson, John Hancock, and Samuel Adams. NFPA 1001: N/A. *FFHB, 2E:* Page 9.

Question #55. Fire extinguishers are labeled to make their firefighting rating quick and easy to identify. The older versions of fire extinguishers are labeled with colored geometrical shapes with letter designations. Newer fire extinguishers are labeled with a picture system. NFPA 1001: 5.3.16. *FFHB, 2E:* Page 188.

Question #56. Specific chemical information is important for a safe and informed response. NFPA 472: 4.2.1 - 4.2.2. *FFHB, 2E:* Page 826.

Question #57. The firefighter should be positioned to prevent being pinned between the hose and the hydrant. At this point, the engine can move forward completing the hose lay. NFPA 1001: 5.3. *FFHB, 2E:* Page 262.

Question #58. Emergency evacuation plans should be drawn in graphic form and posted in each classroom and at various locations throughout the school. NFPA 1001: 5.5.1. *FFHB, 2E:* Page 682-683.

Question #59. Firefighter Carey notices that her co-worker, Firefighter Nally, has become increasingly withdrawn, inattentive, and depressed since returning to the station after a fire. Firefighter Nally might be experiencing critical incident stress. NFPA 1001: 5.3; 6.3. *FFHB, 2E:* Page 738.

Question #60. A pike pole is a commonly used tool in overhaul operations. NFPA 1001: 5.3.13. *FFHB, 2E:* Page 649.

Question #61. For the sake of clarification, NFPA defines structural firefighting as "the activities of rescue, fire suppression, and property conservation in buildings, enclosed structures, aircraft interiors, vehicles, vessels, or like properties that are involved in a fire or emergency situation." NFPA 1001: 5.1.1.2. *FFHB, 2E:* Page 127.

Question #62. First responders who arrive in the first few minutes guide how the remainder of the response will go. NFPA 472: 4.2.1-4.2.3. *FFHB, 2E:* Page 743.

Question #63. When an extension ladder exceeds 40 feet, it is required to be equipped with staypoles. NFPA 1001: 5.3.6; 5.3.9; 5.3.10; 5.3.11; 5.3.12; 5.5.3. *FFHB, 2E:* Page 379.

Question #64. Understanding various protective actions is important for responder health and safety. NFPA 472: 5.2.2 - 5.2.4. *FFHB, 2E:* Page 866.

Question #65. When breaching a wood floor using an axe, locate the floor joists and make cuts close to them. NFPA 1001: 5.3.4; 6.3.2. *FFHB, 2E:* Page 546.

Question #66. Once an overhead door is opened, a tool should be placed (a six-foot hook works well) in the track or a pair of locking pliers should be used on the track to prevent the door from closing. NFPA 1001: 5.3.4; 6.3.2. *FFHB, 2E:* Page 526.

Question #67. The logistics section chief is responsible for securing the facilities, services, equipment, and materials for an incident. NFPA 1001: 6.1.1.1. *FFHB, 2E:* Page 40.

Question #68. The water motor gong, unless connected to another alarm, is a local alarm only. NFPA 1001: 6.5.1. *FFHB, 2E:* Page 331.

Question #69. The dump site is where the water is delivered for quick unloading. Because the tenders arrive already full, it is set up first. NFPA 1001: 5.3.15. *FFHB, 2E:* Page 211.

Question #70. Despite the best efforts of judging the needed length of a hoseline, there will be occasions when the line comes up short and will need to be extended. Whatever the case, all firefighters should be familiar with techniques used to extend their hoselines. NFPA 1001: 5.3.10. *FFHB, 2E:* Page 265.

Question #71. First responders should have a basic awareness of the threat of terrorism and basic response actions. NFPA 472: 4.2.1. *FFHB, 2E:* Page 929.

Question #72. Static pressure is the pressure in the system with no hydrants or water flowing. The pump operator then charges the first line with the desired volume, noting the pressure first. With the flow going, the operator again reads the intake gauge and gets the residual pressure or the remaining pressure left in the system after the flow and friction loss from the flow. The pump operator then compares the percentage of pressure drop from static to residual and determines the amount of additional volumes that may be pumped from that hydrant. NFPA 1001: 5.3.15. *FFHB, 2E:* Page 214.

Question #73. Figure 18-40 If a roof is thick with roof covering, two cuts might have to be made. The first cut is made shallow to remove the covering. NFPA 1001: 5.3.11. *FFHB, 2E:* Page 586.

Question #74. When ready, the rescuer at the head directs the others to "roll patient." NFPA 1001: 4.3. *FFHB, 2E:* Page 479.

Question #75. NFPA 1001: 5.1.1.1. *FFHB, 2E:* Page 238.

Question #76. The key elements of rehab include rest, active cooling, hydration, and nourishment. NFPA 1001: 5.3; 6.3. *FFHB, 2E:* Page 731.

Question #77. A firefighter must be careful not to get caught investigating a smoke alarm activation only to find a CO detector has alarmed to dangerous CO levels. All detector activations should be treated as a worst case scenario. NFPA 1001: 6.5.1. *FFHB, 2E:* Page 313.

Question #78. A firefighter conducting a fire safety inspection finds a business that stores flammable and combustible liquids inside the facility. He knows that quantities less than 5 gallons must be stored in approved safety containers and containers with a capacity greater than 30 gallons may not be stacked. NFPA 1001: 5.5.1. *FFHB, 2E:* Page 669-670.

Question #79. Methane has a vapor density of 0.6. If Methane escapes its container, it can be expected to rise into the air. NFPA 1001: 5.3; 6.3. *FFHB, 2E:* Page 87.

Question #80. Known as hydraulic ventilation, water is employed to create air movement. NFPA 1001: 5.3.11. *FFHB, 2E:* Page 565.

Question #81. The four elements of a fire stream are the pump, water, hose, and nozzle. NFPA 1001: 5.3 NFPA 6.3. *FFHB, 2E:* Page 281.

Question #82. Although similar in many ways to structural PPE, the proximity gear must meet more stringent heat-reflection and wearer insulation standards. NFPA 1001: 5.1.1.2. *FFHB, 2E:* Page 130.

Question #83. In these cases, the firefighter's company officer, incident commander, or incident safety officer may allow firefighters to "dress-down." NFPA 1001: 5.1.1.2. *FFHB, 2E:* Page 137.

Question #84. Understanding potential health effects is important. NFPA 472: 5.2.2 -5.2.3. *FFHB, 2E:* Page 840.

Question #85. An overhand knot is generally used to secure the loose end of the working end after tying a knot. NFPA 1001: 5.1.1.1. *FFHB, 2E:* Page 427.

Question #86. It then naturally follows that the steeper the hill, the faster the fire will travel upward and the slower it will travel downward. NFPA 1001: 5.3.19. *FFHB, 2E:* Page 599.

Question #87. Piercing nozzles were originally designed to penetrate the skin of aircraft and now have been modified to pierce through building walls and floors. NFPA 1001: 5.3 NFPA 6.3. *FFHB, 2E:* Page 286.

Question #88. Firefighters should understand that when heated, as in an automobile fire, these hydraulic fluid-filled bumper systems undergo great stress as the fluid expands. This stress can cause the bumper to be propelled off the car, traveling up to 40 feet or more. Persons standing in front of the bumper (front or rear) when this happens can be severely injured. NFPA 1001: 5.3.7. *FFHB, 2E:* Page 600.

Question #89. The information provided to an emergency communications center by mobile or portable radio must be accurate, clear, and complete. NFPA 1001: 5.2.3. *FFHB, 2E:* Page 67.

Question #90. Is a fog stream the correct choice?If so, which width? Figure 19-19 Firefighters use a fog pattern to shield themselves from radiant heat. NFPA 1001: 5.3.10. *FFHB, 2E:* Page 604.

Question #91. The planning section chief is responsible for the development of the incident action plan. NFPA 1001: 6.1.1.1. *FFHB, 2E:* Page 39.

Question #92. When connecting to a Siamese, the outlet on the far left should be chosen first because this will allow better access for using the spanner wrench to tighten the coupling. Some connections only have clappers installed on the right side. NFPA 1001: 6.5.1. *FFHB, 2E:* Page 324.

Question #93. Immediately exit the hazardous environment. All team members must exit. Never leave a firefighter alone. NFPA 1001: 5.3.1. *FFHB, 2E:* Page 171.

Question #94. Control the fill rate of air to avoid excessive heating or shatter in the cylinder. If the cylinder heats or chatters, reduce the fill rate. NFPA 1001: 5.3.1. *FFHB, 2E:* Page 178.

Question #95. If the extinguisher does not put out the fire, the firefighter should not deploy a second extinguisher. Large fires are beyond the capabilities of portable extinguishers. NFPA 1001: 5.3.16. *FFHB, 2E:* Page 195.

Question #96. Lace-up leather boots that rise well above the ankle (8 to 10 inches) help protect the wearer from cuts, snakebites, and burns. Additionally, a good fitting, tightly laced boot can help prevent ankle sprains and reduce foot fatigue. NFPA 1001: 5.1.1.2. *FFHB, 2E:* Page 132.

Question #97. Understanding various protective actions is important for responder health and safety. NFPA 472: 5.2.2 - 5.2.4. *FFHB, 2E:* Page 874.

Question #98. But be aware there may be additional locking devices out of view that may not be easily unlocked. NFPA 1001: 5.3.4; 6.3.2. *FFHB, 2E:* Page 532.

Question #99. As the use of security gates and overhead doors increases, the power saw has become the tool of choice to remove the door or gate. NFPA 1001: 5.3.4; 6.3.2. *FFHB, 2E:* Page 520.

Question #100. The wrong extinguisher can be worse than no extinguisher. Aiming at other points may cause the fire to spread beyond the capability of the extinguisher. The extinguisher must be aimed from the proper distance, which often is set by the room size and fire size. NFPA 1001: 5.3.16. *FFHB, 2E:* Pages 187 and 195.

PHASE II

Phase Two, Exam Three

1. When speaking with a citizen who is reporting an emergency, the telecommunicator's voice should project authority and knowledge. The telecommunicator can use technical or complex language as long as they maintain a polite and friendly tone.

   a. True

   b. False

2. You are checking for fire extension in an adjacent apartment to the fire apartment. You open doors and windows to allow the smoke to escape. This method is actually preferred to cutting a hole in the roof or the wall.

   a. True

   b. False

3. During a building inspection, it is only necessary to make sure fire extinguishers are in their proper location. It does not matter to the inspector what type of fire extinguisher is present.

   a. True

   b. False

4. If you are in a rural area and decide to use a lake as a water source for a tender shuttle operation, you might make the operation easier if you keep your eyes open for a wall hydrant.

   a. True

   b. False

5. A first responder at the operations level can perform decontamination if they are properly trained and equipped.

   a. True

   b. False

6. You have responded to a call where a propane BBQ cylinder has been stored in a home. The cylinder has ruptured in a closed bedroom. The flammable range of propane will dictate the danger of the situation..

   a. True

   b. False

7. Firefighters' attitudes about safety are affected by the fire department's safety culture, the fire department's history, and the examples set by others.

   a. True

   b. False

8. Class D fires can be fought with any of the available extinguishers.
   a. True
   b. False

9. Because the steel is not required to have significant fire-resistive coatings, Type II buildings are susceptible to collapse.
   a. True
   b. False

10. You are in an airplane hangar and notice it is equipped with a sprinkler system. It is probable that the system is a deluge system.
    a. True
    b. False

11. Firefighter Park is climbing a ladder. She should keep three limbs in contact with the ladder whenever she is moving on a ladder.
    a. True
    b. False

12. Every communications facility should be supported by a backup location in the event that the primary facility encounters problems that render it inoperable and result in evacuation.
    a. True
    b. False

13. A firefighter has raised a ladder to a structure and is determining the proper distance the foot or butt of the ladder should be away from the building. The firefighter can estimate this by moving the foot of the ladder away from the building a distance of one-half the working length of the ladder.
    a. True
    b. False

14. MSDS always provide a high quality of information.
    a. True
    b. False

15. You are filling an aluminum SCBA bottle at the station Cascade unit. You want to ensure you have enough air available to fight a fire. In your estimation, you can perform 35 minutes of work before you will require rehab. You believe you normally use 100 psi per minute; therefore, you should fill your tank to 3,500 psi.
    a. True
    b. False

16. Fitness for duty for firefighters includes mental fitness, physical fitness, wellness, energy, and rest.
    a. True
    b. False

17. Chemtrec can only provide basic chemical information.

    a. True

    b. False

18. For an attack of a grass fire in an interstate median, booster hose, if available, would be an acceptable choice for quick attack.

    a. True

    b. False

19. The best method for drying fire service rope is to use a front loading dryer on high heat.

    a. True

    b. False

20. Hydration is vital for firefighters because of the importance of water to all body systems. Firefighters should wait until they are thirsty to hydrate, allowing the body to dictate when fluids are needed.

    a. True

    b. False

21. All of the senses are the safest form of recognition and identification.

    a. True

    b. False

22. If a firefighter is exposed to a hazardous materials, OSHA issues the regulation that requires the employer provide a physical exam for the firefighter.

    a. True

    b. False

23. Ladder selection can be affected by many things, such as ground condition, purpose, slope of ground, and available personnel.

    a. True

    b. False

24. As you are fighting a propane tank fire, you should remember that the best strategy might be to cool the tank and let the product burn.

    a. True

    b. False

25. When responding to a fire protective system activation, firefighters should go immediately to the actuator panel for information about the activation.

    a. True

    b. False

26. The advantage of fighting a ground cover fire in light fuels is that firefighting operations are much safer.

    a. True

    b. False

27. Setting a knot is the practice of making sure all parts of the knot are lying in the properorientation to the other parts and look exactly as the pictures indicated.

   a. True

   b. False

28. Your partner is a 110-lb. firefighter. She is attempting to use a 16-lb. sledgehammer to breach a wall in training. This is probably not the correct tool for her to use.

   a. True

   b. False

29. As hose is vital to firefighter operations, it is imperative that you remember that hose testing should occur _____.

   a. annually

   b. bi-annually

   c. daily

   d. visually during reloading

30. Your engine has parked at the scene of a structure fire. The driver operator tells you she is going to use a reverse lay. She is referring to _____.

   a. a direct attack

   b. an indirect attack

   c. a master stream

   d. water supply

31. You and your partner are removing a patient from a two-vehicle, high-speed, heavy-damage motor vehicle accident. The passenger SRS device is deployed, but the driver's SRS device has not. While removing the passenger, you remind your partner that _____.

   a. the driver's SRS device is not a concern

   b. the driver's SRS device is still a concern

   c. the SRS devices all deploy simultaneously

   d. SRS devices and motor vehicle accidents are not related

32. The public information officer arrives on the scene of a mass casualty incident where the incident command organizational structure has already been established. To whom will the public information officer reportin the command structure?

   a. The operations section chief

   b. The incident commander

   c. The planning section chief

   d. The finance/administration section chief

33. You are preparing to hoist a pike pole to a second story roof. You are on the ground tying the rope. From your training, you remember that the two ways of tying a clove hitch are _____ or around an object.

    a. in the open

    b. in the air

    c. bowline

    d. Becket bend

34. You are setting up a positive pressure ventilation of a residence filled with smoke. The fire has been put out. You know that the quickest and easiest set up to prepare for overhaul is usually _____ ventilation.

    a. negative pressure

    b. positive pressure

    c. combination

    d. natural

35. You are conducting an inspection of a restaurant kitchen. You see there is a hood extinguisher system. You can suspect the agent is _____.

    a. Class A or B

    b. Class B or C

    c. Class B or K

    d. Class A or K

36. At the operations level, when would the shutting of valves be appropriate?

    a. On rail cars

    b. When they are remote

    c. Only at fuel storage facilities

    d. Always

37. A backpack fire extinguisher and a pressurized water extinguisher _____.

    a. differ in that the backpack extinguisher is shaped to make wildland deployment easier

    b. differ in that the backpack extinguisher is cartridge operated

    c. differ in that the backpack extinguisher is hand operated

    d. have no differences

38. You have stopped the flow of an activated sprinkler head in a hotel at 2 A.M. Only one head activated and there was no fire damage. You have closed off the basement room where the head activated. Your department's next move should be to _____.

    a. allow guests back into their rooms, but ensure no person will enter the basement

    b. close the hotel and not let any person back into the building

    c. require the hotel to set up a fire watch

    d. shut off the water supply to the system to ensure no further accidental activation

39. When inspecting an OS&Y, you are able to tell the valve is closed when
_____.

   a. the indicator says "closed"

   b. the stem is visible

   c. the stem is not visible

   d. there is no way to tell

40. The _____ nozzle was designed for aircraft operations.

   a. fog

   b. piercing

   c. ARFF ram

   d. straight bore

41. If you are working a fireline on a wildland fire, and find yourself in the path of a fast-moving fire, you must ensure you have _____ before deploying your shelter.

   a. a shovel or pulaski available

   b. ear and eye protection

   c. no other options

   d. a radio

42. You are responding to a reported wildland fire incident. Before donning the wildland ensemble, you should ensure _____.

   a. fire-resistive or cotton undergarments are worn

   b. a bandana is tied around your neck

   c. boots are leather with a steel toe

   d. the fire shelter is open and ready to deploy

43. You and two other firefighters are advancing a 2 1/2" hoseline into a structure that has flames and smoke showing. It is a difficult advance as there are many turns in a long hallway. You arrive at the fire and are backing up the firefighter at the nozzle. Your partner behind you taps your shoulder and points to his facemask. He turns to leave as he activates his PASS device and his bypass valve. You should _____.

   a. get your other partner's attention immediately

   b. relay to command that one member of your company is exiting the structure

   c. immediately turn and follow him out of the structure

   d. stop him, use your buddy breathing attachment, and continue backing up the nozzleman

44. Which of the following methods can be used to control external bleeding?

   a. Direct pressure on the site of bleeding

   b. Elevation of the bleeding injury

   c. Using pressure points

   d. All of the above

45. You are forcing entry into a garage. You have managed to open the overhead door. To ensure your safety, the next tool you use should be _____.

    a. a rotary saw

    b. locking pliers

    c. a three-foot pike pole

    d. none of the above

46. You are on scene of a structure fire at a hotel. There is a reported fire in the kitchen of one of the restaurants. During your scene size-up, you find a door that leads to an area where you think you need to be to investigate this fire. The hotel is only 6 months old. Your best access to this area probably will be through _____.

    a. forcing the door with a halligan

    b. cardkey or similar lock

    c. forcing the door with a pick head axe

    d. the front entrance

47. Your rookie firefighter is having trouble with terminology. You remind him that _____ are devices that water flows through.

    a. Higbees

    b. appliances

    c. hose tools

    d. hose bridges

48. An attempt to extinguish a _____ fire with water will result in a violent reaction, releasing heat and brilliant light.

    a. Class A

    b. Class B

    c. Class D

    d. Class K

49. Water sources all have many components. As a tender shuttle operator, you are most concerned with the _____ components.

    a. source and treatment

    b. treatment and distribution

    c. method of storage and treatment

    d. supply and distribution

50. You are doing a monthly maintenance on the SCBA on your engine. You find that the shoulder harness has a half-inch cut in it. You should _____.

    a. log the information on the monthly maintenance sheet

    b. burn the edges of the nylon to keep the cut from tearing more and place the pack back into service

    c. remove the SCBA from service until a technician can replace the damaged parts

    d. do nothing as you are not an SCBA technician

51. You are assisting the driver/operator of a pumper with a water supply operation. The driver/operator wants to draft water from a lake. After positioning the apparatus, she asks you to connect the supply line. You know that you will need to find the

    _____.

    a. large diameter supply line
    b. two 2-1/2-inch hoselines for drafting
    c. hard sleeve and strainer
    d. drafting manual

52. You and your company are on scene of a shed fire at a residence. Your officer tells you to advance a line into the backyard and protect the exposures. Upon entering the backyard, you see the house has a swimming pool. On this scene, you should be most concerned about _____.

    a. accidental drowning
    b. poisonous gas
    c. children remaining in the home
    d. efficient use of water

53. You are working a mutual aid wildland fire in another fire jurisdiction. You hear the Incident Commander call for a tanker. You know the call is for

    _____.

    a. an aircraft
    b. a mobile water supply apparatus
    c. a large fuel truck
    d. you are unsure

54. You are responding to a reported medical problem in a residential home. You enter the home with your work uniform, gloves, goggles, and your med kit. Once inside, you smell natural gas. You should immediately _____.

    a. vacate the IDLH atmosphere
    b. quickly locate the patients and remove them
    c. check the pilot lights on the appliances
    d. triage the patients

55. You are touring a metropolitan airport and your partner notices that there appears to be no fire hydrants. You remind her that this airport most likely utilizes _____.

    a. dry hydrants
    b. a tender shuttle operation
    c. flush-type hydrants
    d. negative gravity flow hydrants

56. As a firefighter, you know that anytime you encounter a PIV, WIV, or OS&Y, it should be _____.

    a. in the open position
    b. in the closed position
    c. locked in the open position
    d. locked in the closed position

57. Perhaps the most important task of using a non-integrated PASS device is ensuring

    _____.

    a. it can be heard outside
    b. it is turned on
    c. it is equipped with extra batteries
    d. the flashing light must be seen through smoke

58. Photoionization (PID) detectors are designed to identify which risk?

    a. Corrosive
    b. Toxic
    c. Radioactive
    d. Explosive

59. You are recalled to the scene of a structure fire. You have been on vacation and have just run a marathon the previous day. When you don your SCBA, it is important to consider that _____.

    a. your legs may no longer support the additional weight of the SCBA
    b. you may have lost weight
    c. you may have forgotten important procedures on your vacation
    d. your partners have gotten used to fighting fire without you

60. You have just finished extinguishing a car fire with a 2½" line. The car was a total loss. The temperature outside is 90° F. You are attempting to open the hood. There is a small amount of steam and/or smoke coming from the front wheel wells. At this point in the fire, you may _____.

    a. remove your SCBA as long as you stay 5 feet from the vehicle
    b. remove your SCBA and coat, even in the vicinity of the vehicle as the hot temperature rules now apply
    c. remove your helmet to allow heat to leave your body as long as you are 5 or more feet from the vehicle
    d. stay in full PPE as long as you are near the vehicle and there is possibly products of combustion

61. To determine how many additional lines can be connected to a fireground operation, it is vital that you remember to check the _____ pressure before any lines are opened.

    a. static
    b. residual
    c. flow
    d. nozzle

62. You are fighting a fire in a commercial warehouse membership club. The fire is in one of the stacks of fertilizer bags. While it is burning, the immediate concern is

    _____.

    a. hazmat
    b. exposure
    c. collapse
    d. smoke

63. Firefighters are responding to a fire in a structure that is Type V construction. The most common occupancy of this type of construction is _____.

    a. residential

    b. commercial

    c. business

    d. educational

64. While on a structure fire, you notice you are having trouble fastening your chin strap on your helmet. You decide that practice sessions of donning your helmet with gloves on at the station would be a good idea, as this will increase _____ and reduce the frustration of lost dexterity.

    a. blood flow

    b. muscle memory

    c. confidence

    d. training hours

65. Nozzle reaction is _____.

    a. the movement of the nozzle opposite from the direction of the water flow

    b. the movement of the nozzle in the same direction as the water flow

    c. dangerous to persons weighing less than the nozzle operator

    d. inversely proportionate to the gpm

66. You and your partner are conducting a search operation in a smoke-filled residential structure. You find two unconscious men in the basement near a pool table. The men are both of equal size and appear to weigh in excess of 200 lbs. Your partner weighs 118 lbs and does not have a lot of upper body strength. You suggest she use a _____ to move the second man to the bottom of the stairs.

    a. foot drag

    b. a webbing sling drag

    c. firefighter's carry

    d. seat carry

67. You have just been issued new work shirts from your quartermaster. You are unsure whether they have to be professionally cleaned or you can wash them yourself. The easiest way to answer your question is _____.

    a. to read the label

    b. to phone the quartermaster

    c. check the Internet

    d. read the proper NFPA standard

68. A 4A rating of an extinguisher should extinguish _____.

    a. 4 square-feet of fire

    b. 4 times the fire of a 1A

    c. 16-square-feet of fire

    d. 4 lbs of class A material

69. The four elements of a fire stream are the _____.

    a. pump, water, hose, and nozzle

    b. source, hose, pressure, and nozzle

    c. psi, gpm, length of hose, and friction loss

    d. type of source, diameter of hose, length of hose, and gallons per minute

70. A backpack fire extinguisher and a pressurized water extinguisher _____.

    a. differ in that the backpack extinguisher is shaped to make wildland deployment easier

    b. differ in that the backpack extinguisher is cartridge operated

    c. differ in that the backpack extinguisher is hand operated

    d. have no differences

71. Hydraulics is the study of fluids _____.

    a. and their expansion capability

    b. and their affect of fire

    c. at rest and in motion

    d. none of the above

72. You are the nozzle operator advancing a hoseline toward an automobile fire. The engine compartment is fully engulfed. You should approach the car from the _____.

    a. rear

    b. front

    c. side

    d. top

73. You are assisting another crew with cleanup and are helping them get their apparatus back in service. The life-safety ropes have been washed and are still wet. Probably the best and one of the easiest methods of drying the ropes is _____.

    a. laying them flat

    b. machine drying them

    c. hanging them

    d. Synthetic ropes require no drying time. They can be bagged and put back on the apparatus.

74. When initially ventilating vertically, sometimes _____ can be initiated before a roof cut is made.

    a. opening a dumbwaiter bulkhead

    b. opening a first floor door

    c. opening a fire floor window

    d. turning on the window air conditioner

75. As a rule of thumb, which materials will burn: organic or inorganic?
    a. Organic
    b. Inorganic
    c. Both are equally likely to burn.
    d. Neither are likely to burn.

76. You are filling an SCBA bottle from the Cascade unit in your station. You notice the bottle is heating rapidly while you are filling it. You should _____.
    a. stop filling the bottle immediately
    b. check the supply air to ensure it is not overheated
    c. reduce the airflow at the bottle
    d. reduce the fill rate

77. When choosing PPE footwear, you should weigh the _____.
    a. cost
    b. advantages
    c. disadvantages
    d. all of the above

78. You and your partner are assigned the task of advancing a hoseline to a second story window. Your partner takes an uncharged hoseline and begins ascent. For a safe operation, you should use _____ as a guide to begin your climb.
    a. the next coupling
    b. the middle of the next section
    c. the third loop
    d. the pre-marked section

79. While bagging life-safety rope, you remind your partner that he should _____ the rope to prevent hang-up.
    a. stuff
    b. coil
    c. wad
    d. fold

80. Which one below is not a requirement to have a hazardous materials physical?
    a. Exposure to a chemical above the PEL
    b. Performing firefighting activities
    c. Member of a HazMat team
    d. Required to wear a respirator

81. The ability of the sun to provide heat to the Earth is an example of what mode of heat transfer?
    a. Conduction
    b. Convection
    c. Radiation
    d. All of the above

82. What fire service innovation was created as a result of numerous fires that destroyed factories during the Industrial Revolution?
    a. Fire alarms
    b. Triple combination engine companies
    c. Fire hydraulics
    d. Automatic sprinkler systems

83. When responding to residential structure fires with sprinkler systems, it is important for firefighters to remember that these systems _____.
    a. are intended to protect vital areas, such as computer rooms or stereo centers
    b. are only intended to protect egress hallways
    c. are intended for life safety and not necessarily property conservation
    d. are intended for property conservation and not necessarily life safety

84. You are attempting to breach a wall. The house you are in is old with lath and plaster walls. You think the wall on which you are working might be a load-bearing wall, and you are not really sure about the structural integrity. You do, however, need to see what is on the other side of the wall, as there is no door to the room beyond. Your best choice would be to cut a _____ shaped hole.
    a. circular
    b. square
    c. triangular
    d. any of the above

85. You find yourself needing to cut a hole in a wooden floor for an emergency escape. You estimate you have about five minutes to leave the area you are in and you have only an axe with which to work. Your first step is to _____.
    a. sound the floor for the joists
    b. cut through the subfloor
    c. request a chainsaw from command
    d. use your PASS

86. You are a tender driver/operator involved in a tender shuttle operation. Your tender holds 1,800 gallons. You must drive 5 minutes to a water source, and it takes 5 minutes to fill your tank and 5 additional minutes to position and dump your tank. If yours is the only tender in operation, you know the maximum fire flow for this fire is _____.
    a. 180 gpm
    b. 90 gpm
    c. 18 gpm
    d. 100 gpm

87. When responding to a fire alarm in a city business, firefighters should don _____ ensemble.
    a. swift-water
    b. structural
    c. wildland
    d. proximity

88. When setting up a tender shuttle operation, you direct that the _____ be set up first.
    a. fill site
    b. dump site
    c. flow site
    d. drafting area

89. For a Class C fire, firefighters should ensure that
    _____.
    a. a Class C only extinguisher is used
    b. CO2 extinguisher is used only if the dry chemical extinguisher is not available
    c. the electrical source is disconnected
    d. a pressurized water extinguisher is used if it contains electrical suppressing foam

90. Out of the following scenarios which is the most likely?
    a. Anthrax attack
    b. Nuclear detonation
    c. Nerve agent attack
    d. Pipe bomb

91. Chief Walter is setting up the organizational structure for a major incident. In what section will he place the staging officer?
    a. Operations
    b. Planning
    c. Logistics
    d. Finance/administration

92. You are an A shift firefighter. At morning pass down, B shift engine company neglects to inform you they used a pressurized-water extinguisher on a mattress fire. You discover this when you go to use the extinguisher on a small trash fire in an office building. This scenario tells you something important. What is it?
    a. B shift company should be written up.
    b. A shift company officer is neglectful.
    c. B shift company officer is neglectful.
    d. A shift morning truck checks were not thoroughly completed.

93. You are responding to a reported medical problem in a residential home. You enter the home with your work uniform, gloves, goggles, and your med kit. Once inside, you smell natural gas. You should immediately _____.
    a. vacate the IDLH atmosphere
    b. quickly locate the patients and remove them
    c. check the pilot lights on the appliances
    d. triage the patients

94. While you are on a firefighting operation, you become thirsty and there is no bottled water available to you. You consider your knowledge of water systems and decide you _____.
    a. can drink water from a city fire hydrant, no matter the system pressure
    b. cannot drink water from a city fire hydrant as it is untreated
    c. can drink water from a city fire hydrant, as long as the system has a residual pressure of 20 psi
    d. can drink water from a city fire hydrant, as long as the system has a static pressure or 20 psi

95. You have been assigned the duty of stopping the flow from an activated sprinkler head. You should _____.
    a. rely on past practice to stop the flow and establish a leak-free seal
    b. use a wedge or sprinkler tongs
    c. prepare to get seriously wet
    d. all of the above

96. A _____ can cause a fire to spread to unaffected areas of the compartment, expanding the amount of area involved.
    a. boiling point
    b. rollover
    c. flash point
    d. heat sink

97. To protect a firefighter's hair, and to offer protection for the straps of a SCBA mask, a firefighter should don the _____.
    a. helmet
    b. protective hood
    c. SCBA
    d. coat

98. During a residential structure fire operation, there is a collapse of the interior first floor. You remind your partner that victims may survive _____.
    a. in the voids
    b. in the basement
    c. in the spans
    d. on the first floor

99. You are advancing a hoseline into a commercial building which is a telemarketing center with many cubicles. As you enter, you look up and notice the ceiling tiles appear to be moving up and down. You decide to _____.
    a. retreat and regroup
    b. unseat the tiles with the hose stream
    c. advance quickly to the seat of the fire
    d. fight the fire from the doorway position

100. You are doing area familiarization in a rural area and notice there are no fire hydrants. It would be prudent for you to ascertain if there are any of these available for your use.

    a. Tanks

    b. Ponds

    c. Cisterns

    d. All of the above

Phase II, Exam III: Answers to Questions

1. F	26. F	51. C	76. D
2. T	27. F	52. B	77. D
3. F	28. T	53. D	78. A
4. F	29. D	54. A	79. A
5. T	30. D	55. C	80. B
6. T	31. B	56. C	81. C
7. T	32. B	57. B	82. D
8. F	33. A	58. B	83. C
9. T	34. B	59. B	84. C
10. T	35. C	60. D	85. A
11. T	36. B	61. A	86. B
12. T	37. C	62. C	87. B
13. F	38. C	63. A	88. B
14. F	39. C	64. B	89. C
15. F	40. B	65. A	90. D
16. T	41. C	66. B	91. A
17. F	42. A	67. A	92. D
18. T	43. A	68. B	93. A
19. T	44. D	69. A	94. A
20. F	45. B	70. C	95. D
21. T	46. B	71. C	96. B
22. T	47. B	72. C	97. B
23. T	48. C	73. C	98. A
24. T	49. D	74. A	99. B
25. F	50. C	75. A	100. D

Phase II, Exam III:
Rationale & References for Questions

Question #1. When speaking with a citizen who is reporting an emergency, the telecommunicator's voice should project authority and knowledge. The telecommunicator should use plain, everyday language at all times when speaking to the public and maintain a polite and friendly tone. NFPA 1001: 5.2.1; 5.2.2. *FFHB, 2E:* Page 52.

Question #2. This choice is the most appropriate for incidents where time is not essential and there is no urgency to the emergency, permitting a slower venting operation. NFPA 1001: 5.3.11. *FFHB, 2E:* Page 563.

Question #3. Firefighters should inspect for the presence of fire extinguishers and ensure that the extinguisher has the proper classification and rating for its location. NFPA 1001: 5.5.1. *FFHB, 2E:* Page 666.

Question #4. A dry hydrant is not really a fire hydrant but a connection point for drafting from a static water source such as a pond or stream, not a pressurized one. They are used primarily in rural areas with no water system. They may be found in urban and suburban areas as a backup where certain buildings may be of great distance from other water sources, but a pond or stream is located nearby. NFPA 1001: 5.3.15. *FFHB, 2E:* Page 208.

Question #5. Understanding various protective actions is important for responder health and safety. NFPA 472:5.4.1. *FFHB, 2E:* Page 877.

Question #6. This range will determine whether the fuel is too rich or too lean to burn in the given space it occupies. If the spill or leak is inside a structure with no windows or doors open, then there is a chance that the gas is too thick or rich to burn, as the ratio between oxygen and the gas will not support combustion. NFPA 1001: 6.3.1. *FFHB, 2E:* Page 602.

Question #7. Firefighters' attitudes about safety are affected by the fire department's safety culture, the fire department's history, and the examples set by others. NFPA 1001: 6.1.1.1. *FFHB, 2E:* Page 116.

Question #8. Great care must be used when attempting to extinguish a fire in these types of fuels. Water and other extinguishing agents can react violently when applied to burning combustible metals and can endanger nearby firefighters. Also, because of the differences in the metal and alloy fuels, there is no universal Class D extinguishing agent that works on all Class D materials. NFPA 1001: 5.3.16. *FFHB, 2E:* Page 186.

Question #9. Because the steel is not required to have significant fire-resistive coatings, Type II buildings are susceptible to collapse. NFPA 1001: 5.3.10; 6.3.2. *FFHB, 2E:* Page 354.

Question #10. Deluge systems are designed to protect areas that may have a fast-spreading fire that could engulf an entire area. Petroleum-handling facilities, aircraft hangars, some manufacturing facilities, and hazardous materials storage areas are all examples of occupancies that may have a deluge system. NFPA 1001: 6.5.1. *FFHB, 2E:* Page 320.

Question #11. Firefighter Park is climbing a ladder. She should keep three limbs in contact with the ladder whenever she is moving on a ladder. NFPA 1001: 5.3.6; 5.3.9; 5.3.10; 5.3.11; 5.3.12. *FFHB, 2E:* Page 383.

Question #12. Every communications facility should be supported by a backup location in the event that the primary facility encounters problems that render it inoperable and result in evacuation. NFPA 1001: 5.2.1. *FFHB, 2E:* Page 50.

Question #13. A firefighter has raised a ladder to a structure and is determining the proper distance the foot or butt of the ladder should be away from the building. The firefighter can estimate this by moving the foot of the ladder away from the building a distance of one-quarter the working length of the ladder. NFPA 1001: 5.3.6; 5.3.9; 5.3.10; 5.3.11; 5.3.12. *FFHB, 2E:* Page 400.

Question #14. Specific chemical information is important for a safe and informed response. NFPA 472: 4.2.1 - 4.2.2. *FFHB, 2E:* Page 818.

Question #15. Aluminum cylinders are used for thirty-minute (2216-psi) rated SCBA units. NFPA 1001: 5.3.1. *FFHB, 2E:* Page 154.

Question #16. Fitness for duty for firefighters includes mental fitness, physical fitness, wellness, energy, and rest. NFPA 1001: 5.3; 6.3. *FFHB, 2E:* Page 726.

Question #17. Specific chemical information is important for a safe and informed response. NFPA 472: 4.2.1 - 4.2.2. *FFHB, 2E:* Page 827.

Question #18. Booster hose is smaller-diameter, rubber-coated hose of 3/4 or 1-inch size usually mounted on reel that can be used for outside fires or overhaul operations after the fire is out. Booster lines have a limited flow rate of up to 30 gpm and, although easy to maneuver and control, they are unsafe for structural firefighting. NFPA 1001: 5.3.15. *FFHB, 2E:* Page 222.

Question #19. Hanging to dry. This is one of the eaiser methods of drying ropes. NFPA 1001: 5.3. *FFHB, 2E:* Page 443.

Question #20. Hydration of firefighters should become paramount, even to the point of excess. Firefighters must not wait until they are thirsty to drink water! NFPA 1001: 5.3; 6.3. *FFHB, 2E:* Page 732.

Question #21. NFPA 472: 5.2.2 - 5.2.4. *FFHB, 2E:* Page 755.

Question #22. First responders should understand laws, regulations, and standards. NFPA 472: 4.2.1. *FFHB, 2E:* Page 748.

Question #23. Ladder selection can be affected by many things, such as ground condition, purpose, slope of ground, and available personnel. NFPA 1001: 5.3.6; 5.3.9; 5.3.10; 5.3.11; 5.3.12. *FFHB, 2E:* Page 385.

Question #24. 3. For instance, get water on exposed tanks to reduce internal pressure buildup and prevent the possibility of a BLEVE. 4c. For some fires, consider letting the fire burn itself out. NFPA 1001: 5.3.10. *FFHB, 2E:* Page 626.

Question #25. In newer systems, the protective systems are integrated into a single "smart" system that includes an annunciator panel, fire alarm control panel, and system override controls. NFPA 1001: 6.5.1. *FFHB, 2E:* Page 314.

Question #26. In fact, this deceptiveness has caused a great number of firefighter deaths and injuries over the years. Numbers of firefighters have been overrun in grassy fuels thinking they could "outrun" the fire when things went bad. NFPA 1001: 5.3.19. *FFHB, 2E:* Page 597.

Question #27. Dressing a knot is the practice of making sure all parts of the knot are lying in the properorientation to the other parts and look exactly as the pictures indicated. NFPA 1001: 5.1.1.1. *FFHB, 2E:* Page 426.

Question #28. A tool that is too heavy cannot be moved fast enough to develop proper force. NFPA 1001: 5.3.4; 6.3.2. *FFHB, 2E:* Page 513.

Question #29. The testing of hose begins with a visual inspection of the hose coupling. This inspection should be done with the annual test and during routine reloading and reconnection of hose sections. As the hose is loaded, it should also be visually inspected for any type of damage. NFPA 1001: 6.5.3. *FFHB, 2E:* Page 274.

Question #30. A reverse lay is the opposite of the forward lay with the supply line being dropped off at the fire location and the engine laying the hose toward the water source. NFPA 1001: 5.3.10. *FFHB, 2E:* Page 270.

Question #31. SRS devices that have not deployed should be considered "live." NFPA 1001: 6.4.1. *FFHB, 2E:* Page 488.

Question #32. The public information officer is a command staff position and reports directly to the incident commander. NFPA 1001: 6.1.1.1. *FFHB, 2E:* Page 41.

Question #33. NFPA 1001: 5.1.1.1. *FFHB, 2E:* Page 238.

Question #34. A distinct advantage of the PPV blower is ease of setup. If properly applied, the blower can do the job of several firefighters attempting to open multiple holes with ladders at different locations. NFPA 1001: 5.3.11. *FFHB, 2E:* Page 565.

Question #35. Its fuels are similar to Class B fuels, but involve high-temperature cooking oils and, therefore, have special characteristics. Typically, firefighters have used Class B extinguishers on these types of fires, but they have been less effective on deep layers of cooking oils. NFPA 1001: 5.3.16. *FFHB, 2E:* Page 187.

Question #36. Responders should understand the protective actions that are available to utilize. NFPA 472: 5.3.2 - 5.3.4. *FFHB, 2E:* Page 898.

Question #37. Pump-type extinguishers are hand-pumped devices of two designs, depending on whether the pump is internal or external to the tank. Pressurized-water, pressurized-loaded stream, and stored-pressure extinguishers operate by means of an expelling gas that propels the agent out of the container. NFPA 1001: 5.3.16. *FFHB, 2E:* Page 188.

Question #38. In these cases, the fire department should not allow the building to be occupied without an approved fire watch system or sprinkler restoration. NFPA 1001: 6.5.1. *FFHB, 2E:* Page 329.

Question #39. When the stem is exposed or outside, the valve is open. NFPA 1001: 6.5.1. *FFHB, 2E:* Page 324.

Question #40. Piercing nozzles were originally designed to penetrate the skin of aircraft and now have been modified to pierce through building walls and floors. NFPA 1001: 5.3 NFPA 6.3. *FFHB, 2E:* Page 286.

Question #41. A fire shelter is a last-resort protective device for firefighters caught or trapped in an environment where a firestorm or blowup is imminent. NFPA 1001: 1977. *FFHB, 2E:* Page 132.

Question #42. The wildland PPE ensemble is designed to be worn over undergarments. These undergarments (long-sleeve t-shirt, pants, and socks) should be 100 percent cotton or of a fire-resistive material. NFPA 1001: 5.1.1.2. *FFHB, 2E:* Page 130.

Question #43. Immediately exit the hazardous environment. All team members must exit. Never leave a firefighter alone. NFPA 1001: 5.3.1. *FFHB, 2E:* Page 171.

Question #44. All these methods can be used to control external bleeding. NFPA 1001: 4.3. *FFHB, 2E:* Page 711-712.

Question #45. Once an overhead door is opened, a tool should be placed (a six-foot hook works well) in the track or a pair of locking pliers should be used on the track to prevent the door from closing. NFPA 1001: 5.3.4; 6.3.2. *FFHB, 2E:* Page 526.

Question #46. Streetsmart Tip Many office occupancies and hotels/motels are using electronic locks for security. The fire department should contact the facility management people to arrange a procedure for obtaining a master card key if available. NFPA 5.3.4; 6.3.2. *FFHB, 2E:* Page 530.

Question #47. Appliances are devices that water flows through, including adapters and connectors. NFPA 1001: 5.3.15. *FFHB, 2E:* Page 227.

Question #48. An attempt to extinguish a Class D fire with water will result in a violent reaction, releasing heat and brilliant light. NFPA 1001: 5.3; 6.3. *FFHB, 2E:* Page 101.

Question #49. The important parts for firefighters are the supply and distribution system, including storage facilities. NFPA 1001: 5.3.15. *FFHB, 2E:* Page 205.

Question #50. Any irregularities should be noted and repaired, or the SCBA should be pulled from service until a department technician can repair the unit. NFPA 1001: 5.3.2. *FFHB, 2E:* Page 172.

Question #51. The engine has to be able to position itself close enough to place the hard sleeve and strainer into the water or else reach the dry hydrant location. NFPA 1001: 5.3. *FFHB, 2E:* Page 221.

Question #52. A swimming pool at a residential occupancy indicates storage of chemicals such as chlorine, which will produce a poisonous gas. Firefighters must understand that there is no "routine" fire. Even a light smoke condition at this type of structure could be deadly. NFPA 1001: 5.3.1. *FFHB, 2E:* Page 146.

Question #53. The term water tender is used to describe mobile water supply apparatus, although some jurisdictions use the term tanker. The fire service is still divided on this language to some degree. The guide refers to a tanker as an aircraft capable of carrying and dropping water for firefighting operations. NFPA 1001: 5.3.15. *FFHB, 2E:* Pages 204-205.

Question #54. A work uniform that meets NFPA standards is not designed to protect the wearer from IDLH atmospheres, but can add another layer of reasonable protection under wildland, proximity, and structural ensembles. NFPA 1001: 5.1.1.2. *FFHB, 2E:* Page 136.

Question #55. The purpose of mounting the hydrant below grade allows access to a water source in areas where an above-grade hydrant would interfere with operation of the facility. Airports, shipping terminals, and many European cities utilize flush-mounted hydrants. NFPA 1001: 5.3.15. *FFHB, 2E:* Page 209.

Question #56. These valves must have a chain lock on them to prevent tampering. A wrench or a wheel controls these valves; a padlock and chain are used to lock them open. NFPA 1001: 6.5.1. *FFHB, 2E:* Pages 324 and 325.

Question #57. The biggest problem with PASS devices results when wearers simply forget to turn their units on. This simple mental lapse has contributed to numerous firefighter fatalities. NFPA 1001: 5.1.1.2. *FFHB, 2E:* Page 135.

Question #58. Responders should understand the protective actions that are available to utilize. NFPA 472: 6.2.1 - 6.2.2. *FFHB, 2E:* Page 905.

Question #59. Even though face pieces have been fitted and tested, weight loss or a 24-hour growth of facial hair may affect the ability to obtain a good seal. NFPA 1001: 5.3.1. *FFHB, 2E:* Page 150.

Question #60. Common sense dictates that firefighters should use SCBA on every fire scene—from start to finish. NFPA 1001: 5.3.1. *FFHB, 2E:* Page 148.

Question #61. Static pressure is the pressure in the system with no hydrants or water flowing. The pump operator then charges the first line with the desired volume, noting the pressure first. With the flow going, the operator again reads the intake gauge and gets the residual pressure or the remaining pressure left in the system after the flow and friction loss from the flow. The pump operator then compares the percentage of pressure drop from static to residual and determines the amount of additional volumes that may be pumped from that hydrant. NFPA 1001: 5.3.15. *FFHB, 2E:* Page 214.

Question #62. Piled storage is held up by the integrity of the items stored, and as that changes with either fire or water damage, those piles may topple. NFPA 1001: 5.3.10. *FFHB, 2E:* Page 627.

Question #63. Firefighters are responding to a fire in a structure that is Type V construction. The most common occupancy of this type of construction is residential. NFPA 1001: 5.3.10; 6.3.2. *FFHB, 2E:* Page 360.

Question #64. The simple act of practicing hands-on tool use with gloves on can increase muscle memory and reduce the frustration of lost dexterity. NFPA 1001: 5.1.1.2. *FFHB, 2E:* Page 129.

Question #65. Nozzle reaction is the force of nature that makes the nozzle move in the opposite direction of the water flow. NFPA 1001: 5.3 NFPA 6.3. *FFHB, 2E:* Page 282.

Question #66. The webbing drag enables a rescuer who is significantly smaller than the victim to perform a rescue. NFPA 1001: 5.3.9. *FFHB, 2E:* Page 472.

Question #67. The key to maintaining personal protective equipment in a high state of readiness is simple: Follow the specific instructions given by the manufacturer. NFPA 1001: 5.1.1.2. *FFHB, 2E:* Page 136.

Question #68. For instance, a 2-A extinguisher will put out twice the fire of a 1-A. NFPA 1001: 5.3.16. *FFHB, 2E:* Page 194.

Question #69. The four elements of a fire stream are the pump, water, hose, and nozzle. NFPA 1001: 5.3 NFPA 6.3. *FFHB, 2E:* Page 281.

Question #70. Pump-type extinguishers are hand-pumped devices of two designs, depending on whether the pump is internal or external to the tank. Pressurized-water, pressurized-loaded stream, and stored-pressure extinguishers operate by means of an expelling gas that propels the agent out of the container. NFPA 1001: 5.3.16. *FFHB, 2E:* Page 188.

Question #71. Hydraulics is the study of fluids at rest and in motion, which describes the flow pattern of water supply and fire streams. NFPA 1001: 5.3 NFPA 6.3. *FFHB, 2E:* Page 290.

Question #72. Safety Firefighters should understand that when heated, as in an automobile fire, these hydraulic fluid-filled bumper systems undergo great stress as the fluid expands. This stress can cause the bumper to be propelled off the car, traveling up to 40 feet or more. Persons standing in front of the bumper (front or rear) when this happens can be severely injured. NFPA 1001: 5.3.7. *FFHB, 2E:* Page 600.

Question #73. Hanging to Dry This is one of the easier methods of drying ropes. NFPA 1001: 5.3. *FFHB, 2E:* Page 443.

Question #74. For example, opening a bulkhead door will be quick and effective. It should be performed before a roof cut is initiated. NFPA 1001: 5.3.11. *FFHB, 2E:* Page 576.

Question #75. As a rule of thumb, only organic materials will burn. NFPA 1001: 5.3; 6.3. *FFHB, 2E:* Page 80.

Question #76. Control the fill rate of air to avoid excessive heating or shatter in the cylinder. If the cylinder heats or chatters, reduce the fill rate. NFPA 1001: 5.3.1. *FFHB, 2E:* Page 178.

Question #77. Now firefighters can choose from traditional rubber-like boots or leather-type boots. Each has advantages and disadvantages. NFPA 1001: 5.1.1.2. *FFHB, 2E:* Page 129.

Question #78. The next firefighter places the hoseline over the left shoulder at the next coupling and begins to climb the ladder. NFPA 1001: 5.3. *FFHB, 2E:* Page 259.

Question #79. Do not coil the rope in the bag; it will hang-up almost every time if you do. NFPA 1001: 5.1.1.1. *FFHB, 2E:* Page 447.

Question #80. First responders should understand laws, regulations, and standards. NFPA 472: 4.2.1. *FFHB, 2E:* Page 748.

Question #81. The ability of the sun to provide heat to the Earth is an example of radiation. NFPA 1001: 5.3; 6.3. *FFHB, 2E:* Page 97.

Question #82. Henry Parmalee invented the automatic sprinkler system to help prevent the major fire losses experienced in factories during the Industrial Revolution. NFPA 1001: N/A. *FFHB, 2E:* Page 14.

Question #83. Residential sprinklers are designed for life safety and not necessarily to protect property. NFPA 1001: 6.5.1. *FFHB, 2E:* Page 314.

Question #84. Streetsmart Tip When making a large opening, a triangular or diamond-shaped hole will help maintain the structural integrity of the wall. NFPA 1001: 5.3.4; 6.3.2. *FFHB, 2E:* Page 545.

Question #85. When breaching a wood floor using an axe, locate the floor joists and make cuts close to them. NFPA 1001: 5.3.4; 6.3.2. *FFHB, 2E:* Page 546.

Question #86. 1800 gals/ 5 mins (drive time from source) + 5 mins (dump time) + 5 minutes (drive time to source) + 5 minutes (fill time) = 90 gallons per minute. NFPA 1001: 5.3.15. *FFHB, 2E:* Page 212.

Question #87. For the sake of clarification, NFPA defines structural firefighting as "the activities of rescue, fire suppression, and property conservation in buildings, enclosed structures, aircraft interiors, vehicles, vessels, or like properties that are involved in a fire or emergency situation." NFPA 1001: 5.1.1.2. *FFHB, 2E:* Page 127.

Question #88. The dump site is where the water is delivered for quick unloading. Because the tenders arrive already full, it is set up first. NFPA 1001: 5.3.15. *FFHB, 2E:* Page 211.

Question #89. The recommended method of fighting these fires is to turn off or disconnect the electrical power and then use an appropriate extinguisher, depending on the remaining fuel source.Class C only extinguishers are not made. NFPA 1001: 5.3.16. *FFHB, 2E:* Page 186.

Question #90. First responders should have a basic awareness of the threat of terrorism and basic response actions. NFPA 472: 4.2.1. *FFHB, 2E:* Page 932.

Question #91. Staging is part of the operations section under the direct control of the operations chief with the assistance of the staging area manager. NFPA 1001: 6.1.1.1. *FFHB, 2E:* Page 39.

Question #92. Extinguishers in buildings should be checked every 30 days, and extinguishers on apparatus should be inspected each time the vehicle is inspected. NFPA 1001: 5.3.16. *FFHB, 2E:* Page 198.

Question #93. A work uniform that meets NFPA standards is not designed to protect the wearer from IDLH atmospheres, but can add another layer of reasonable protection under wildland, proximity, and structural ensembles. NFPA 1001: 5.1.1.2. *FFHB, 2E:* Page 136.

Question #94. After treatment for drinking, water goes into the distribution system or water mains. The mains are divided into feeders and distribution lines. Primary feeders supply secondary feeders and then distribution lines. Distributor lines are the water mains with the building connections and fire hydrants, and for fire protection; they range in size from 6 to 16 inches. NFPA 1001: 5.3.15. *FFHB, 2E:* Page 205.

Question #95. While it may seem easy, stopping water flow from a sprinkler head requires practice to effectively stop the flow and establish a leak-free seal. The firefighter assigned to stop sprinkler flow at the head will get seriously wet. NFPA 1001: 6.5.1. *FFHB, 2E:* Page 326.

Question #96. A rollover can cause a fire to spread to unaffected areas of the compartment, expanding the amount of area involved. NFPA 1001: 5.3; 6.3. *FFHB, 2E:* Page 99.

Question #97. Hoods are made of fire-resistive, form-fitting cloth that protects the face, ears, hair, and neck in areas not covered by the helmet, earflaps, and coat collar. NFPA 1001: 5.1.1.2. *FFHB, 2E:* Page 129.

Question #98. Voids spaces within a collapsed area that are open and may be an area where someone could survive a building collapse. NFPA 1001: 6.4.2. *FFHB, 2E:* Page 506.

Question #99. If the fire is already in the space, the tiles can usually be unseated with the hoseline merely by directing the stream upward and extinguishing fire in the ceiling while advancing. NFPA 1001: 5.3.12. *FFHB, 2E:* Page 587.

Question #100. An additional source of water can come from a developed source of water such as water tanks, ponds, and cisterns. NFPA 1001: 5.3.15. *FFHB, 2E:* Page 205.

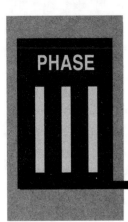

SYNTHESIS & EVALUATION

Referring to Table I-1, the final levels of Bloom's Taxonomy, Cognitive Domain, are covered in section three. Mastery of this section suggests the highest level of understanding of the material. The levels addressed are:

- synthesis
- evaluation

The successful test-taker should be able to rearrange material, modify processes, compare data, and interpret results. For testing at this level, questions will be tied around more of an application process. The student should be able to apply the information learned in the class or in the textbook.

Phase Three, Exam One

1. As you advance up the stairwell of a multistory residential building with a fire alarm, you know that you will need hose to connect to the standpipe. You can use either hose from your engine or the hose stored in the cabinet.

    a. True

    b. False

2. Any action designed to break the accident chain is known as an intervention. Intervention is typically a proactive action.

    a. True

    b. False

3. You are the second-due engine on the scene of a motor home fire in a very crowded parking lot of a commercial grocery store. The vehicle is about 50 feet from the outdoor garden center of the store. Your engine was on a medical call nearby and was clearing scene when the officer on the first-due engine requested your assistance. An attack line has been pulled and the attack crew is beginning to fight the fire. IC tells you to set up exposure protection. You should immediately begin cooling down the garden shop with a second 2½" attack line.

    a. True

    b. False

4. An engine company arrives on the scene of a large storage building that is fully involved in flame with heavy smoke conditions. Bystanders in the next building report that the shed on fire is unoccupied and very little remained in storage in the facility. Using the concept of risk/benefit in this situation, you can assume that the team will make entry into the structure for search and rescue and an aggressive fire attack.

    a. True

    b. False

5. In selecting the correct ladder for a structure, firefighters can use estimates of normal heights of floors for residential and commercial structures. Residences measure approximately 8-10 feet from floor to floor. The average commercial story will average approximately 10-12 feet from floor to floor.

    a. True

    b. False

6. You are a ventilation group supervisor and your job it to ventilate the roof of a four-story residential structure. Before you ascend the ladder, you should brief your crews of the location of the second roof exit.

    a. True

    b. False

7. You are on the scene of a fire in the mainframe computer room of a business. The computer overheated and was extinguished by a part-time employee using a 2A10BC extinguisher. Had you been the first on scene, this extinguisher would have been your first choice of extinguisher to use, also.
   a. True
   b. False

8. When you tactilely inspect a rope, you are visually inspecting the rope.
   a. True
   b. False

9. Grasping the back plate and lifting the back plate/cylinder overhead to don an SCBA is called the coat method.
   a. True
   b. False

10. You are tasked with making a supply line connection to your engine. You choose not to use a hard sleeve connection because it takes additional manpower and requires the engine to park precisely.
    a. True
    b. False

11. You are fighting a room and contents fire in a residence. You are unsure if the room has been evacuated. You use a combination attack with coordinated ventilation to increase any victim chance of survival. This is a correct course of action.
    a. True
    b. False

12. A fire inspector from the Delmar Fire Department is conducting a fire prevention inspection in a new business. When he arrives, the business owner does not allow him to enter the building or conduct his inspection as planned. He will have to obtain an administrative warrant to enter the building without the consent of the business owner.
    a. True
    b. False

13. It is not necessary to know the identification of the material that is leaking before you start to knock down vapors.
    a. True
    b. False

14. You are directed by your IC to conduct a primary search. On your way to the entry door, you take a moment to look into an uninvolved room's window to check for occupants. This is considered part of the primary search.
    a. True
    b. False

15. The risk/benefit analysis provides that when a little is to be gained, a lot of risk is taken.
    a. True
    b. False

16. As a building burns, the structural elements decompose and lose their strength. This causes a change in the forces and the way the design loads are applied, leading to structural failure and collapse.

    a. True

    b. False

17. While inspecting a business, you look at the fire extinguisher and see a large dent in it. It looks like it still has a charge on it, and does not appear to be compromised. You should assume the extinguisher is operable as it has a current tag on it.

    a. True

    b. False

18. You are in staging and watching the scene of a structure fire. You see one of the attack lines suddenly jump off the ground. You can suspect a water hammer.

    a. True

    b. False

19. Although the incident commander maintains the responsibility for safe control of an incident, he or she may delegate authority for a function or task, transferring the responsibility for safe control of that function or task to the subordinate.

    a. True

    b. False

20. Surface-to-mass ratio is a variable that can determine the amount of time a material can resist gravity and fire degradation. A material with a large surface area and small mass will be slower to burn or fail.

    a. True

    b. False

21. The American fire helmet was developed by a leather hat maker named Andrew Gratacap. The original hats were shaped like top hats and had the name of the fire company painted on them. Later, the idea of a short front brim and an extended back brim was incorporated into the design to lighten the overall helmet and better distribute the weight.

    a. True

    b. False

22. Upon your arrival at a high-rise fire alarm, IC instructs you to find the location of the alarm activation. You and your crew should proceed to the emergency control center to find the location of sprinkler head activation

    a. True

    b. False

23. By-products of combustion, such as toxic chemicals, particulate matter, and heat, can cause damage to and diseases of the lung tissue. The use of protective breathing apparatus in fires is required because of this danger.

    a. True

    b. False

24. Atoms that are lacking electrons are less likely to link up and form molecules than atoms that have a satisfactory electron/proton ratio.

    a. True

    b. False

25. When firefighters are participating in salvage operations, it is always best to cover items to be protected rather than moving items.

    a. True

    b. False

26. Two firefighters come to the aid of a small child who was injured in a bicycle accident. Because the patient is a small child, it is acceptable for the firefighters not to wear BSI equipment, like gloves, to avoid scaring the child.

    a. True

    b. False

27. If you are in a rural area and decide to use a lake as a water source for a tender shuttle operation, you might make the operation easier if you keep your eyes open for a dry hydrant.

    a. True

    b. False

28. Telecommunicators must be able to prioritize incoming lines in order to ensure that the most important call gets the fastest attention. Incoming telephone calls should be answered in the following priority: 1. direct lines 2. emergency or 9-1-1 lines 3. business or administrative lines

    a. True

    b. False

29. Firefighters arrive on a scene to find a Class C fire. Which of the following methods should the firefighters use to best control this type of fire?

    a. Cooling the fire with large quantities of water

    b. The application of a smothering agent to prevent oxygen from getting to the fuel

    c. Removing the flow of electricity and then re-analyzing the class of the fire for further actions

    d. Applying a dry powder to stop the chemical reaction

30. You have just received your new wildland PPE ensemble. As you are unpacking it, you realize you have questions about the care and limitations of use of the helmet. You should_____.

    a. check NFPA 1001

    b. check the information that came with the PPE

    c. check the tag inside the helmet

    d. ask your officer

31. You are in staging on a hazardous materials call. The hazmat technicians are suiting up for entry into a commercial structure in which you used to work. In your estimation, you calculate that the room they need to access is about 5 minutes into the building. You see them donning open-circuit supplied air respirators. You know the cylinders they will be carrying only contain 5 to 10 minutes of air. You should _____.

    a. immediately notify IC

    b. immediately notify staging

    c. start the timer on your watch and monitor the situation

    d. do nothing

32. What is a combustible gas detector designed to read?

    a. Lower flash points

    b. Lower explosive level

    c. Toxic and corrosive gases

    d. Upper flash point

33. You are on scene of a car fire. You find the vehicle to be a diesel pickup truck, so you decide to use foam. You know that diesel is a _____ and this will determine the type of foam you will use.

    a. bio fuel

    b. hydrocarbon

    c. polar solvent

    d. Class A fuel

34. What is the great danger in fighting a fire involving nuclear fuel?

    a. Toxic chemicals

    b. Ineffectiveness of water in combating the fire

    c. Exposure to radiation

    d. Heat

35. You are preparing to hoist a ladder to the third floor of a commercial building. You remember that you will use the _____ of the rope for work such as this.

    a. working end

    b. standing end

    c. standing part

    d. running end

36. As an active firefighter, you should realize that PPE provides you the _____ level of protection and should be considered the _____ of protection.

    a. minimum / last resort

    b. maximum / front line

    c. minimum / front line

    d. maximum / last resort

37. You have responded to a residential structure fire in a neighborhood that is known for its high crime rates. It is 2 A.M. and there is a car parked in front of the home on the street. The home has bars over the windows. You are on the search team and you attempt to force the front door for entry. You spend a minute or so attempting to force the door and it simply will not move. You _____.
    a. suspect an auxiliary locking device
    b. should keep trying to force the door
    c. should try pulling the lock
    d. should try using a K tool

38. You respond to a fuel station with a propane pump burning. Your first action should be to _____.
    a. pull a hoseline
    b. grab a Class B extinguisher
    c. move the closest vehicle away
    d. shut off or have the pumps shut off

39. Two firefighters are performing overhaul operations inside a structure. Which of the following sets of tools would be most useful to the firefighters?
    a. Salvage covers, a hammer, and nails
    b. Bolt cutters and sprinkler wedges
    c. Squeegees and a water vacuum
    d. A carry-all, pike poles, and axes

40. You are called to a boat fire on a lake. Upon arrival, you find a large wooden boat burning.The boat is docked. There are no immediate fire hydrants visible. You should _____.
    a. use a reverse lay to the nearest hydrant
    b. request a water tender
    c. use booster tank water sparingly
    d. consider drafting

41. You are in mop-up mode on a structure fire where you used a considerable amount of foam. The driver/operator of the engine tells you the foam system on the apparatus uses an around-the-pump proportioner. You should expect to _____.
    a. deplete the entire foam capacity
    b. back-flush the pump
    c. prime the pump
    d. discard the first section of hose

42. Once you don your face piece, you should check for a proper seal by _____.
    a. placing your hand over the regulator hole and attempting to inhale
    b. attaching the regulator, taking a breath and holding it, and listening and feeling for any air leaks
    c. smelling for smoke or fumes once you enter the IDLH
    d. No check is needed. It is mandated by NFPA that all masks be custom fit.

43. Which of the following is not a consideration when folding salvage covers?

    a. The deployment method for the salvage cover

    b. The amount of space in the storage compartment where the salvage cover is located

    c. Neat and orderly folds

    d. The size of the salvage cover

44. An engine company is out conducting fire safety inspections in a row of businesses in their first due area. Firefighter Jackson is going to perform her inspection from left to right inside the building. Firefighter Wheatley says that he is going to perform his inspection from outside to inside, and then from top to bottom. Who is correct?

    a. Firefighter Jackson

    b. Firefighter Wheatley

    c. Both are correct.

    d. Neither are correct.

45. You are a firefighter assigned to the airport station. You respond to a reported in-flight emergency that might result in a fiery crash. En route to the staging point, you realize you have your structural helmet instead of your proximity helmet. Your concern would be _____.

    a. radiant facial burns from a melting shield

    b. your helmet is black and not silver

    c. less impact protection from special grade aircraft structural members

    d. "ARFF blindness" due to special coatings not being applied to your structural face shield

46. You and your partner have exited a residential structure fire after conducting a primary search. You return to your engine to change out your bottles and you notice your partner is quite pale. What should you do?

    a. Tell her what you see and encourage her to drink some water or Gatorade.

    b. Tell her what you see and have her sit in the shade for a few minutes while drinking fluid replacement drinks.

    c. Tell her what you see and suggest she report to rehab.

    d. Assume she knows her ability. If she does not complain, then it is safe to assume she is fine and ready for another assignment.

47. When are firefighters and EMS providers exposed to hazardous materials?

    a. Very rarely

    b. Every day in fire suppression and EMS duties

    c. Never

    d. Only when in a factory

48. As a tender shuttle driver/operator, you should pick a dump site that has _____.

    a. access

    b. turnaround area

    c. safety

    d. all of the above

49. A firefighter is working on a fitness and wellness program for the firefighters on her shift. Which of the following elements would be appropriate for consideration in her program?

    a. Cardiovascular conditioning

    b. Core strengthening

    c. Flexibility improvement

    d. All of the above

50. When overhauling a vehicle involved in a car fire, your SCBA fails. The face mask "sucks to your face." The purge valve does not remedy the situation and the bottle is turned full on. As the fire is extinguished and the smoke has dissipated, your course of action should be to _____.

    a. retreat

    b. remove the mask and continue to search for clues pointing to the origin of the fire

    c. remove the mask and move to the second position on the hoseline

    d. ask command to send you a new airpack

51. Firefighters arrive on a scene to find a Class A fire. Which of the following methods should the firefighters use to best control this type of fire?

    a. Cooling the fire with large quantities of water

    b. Applying a smothering agent to prevent oxygen from getting to the fuel

    c. Removing the flow of electricity and then re-analyzing the class of the fire for further actions

    d. Applying a dry powder to stop the chemical reaction.

52. You are a new firefighter assigned to a station that handles technical rescue for the fire district. You would like to customize your structural trousers with sewn-in webbing. You can have this customizing done _____.

    a. at any tent repair store or other store that deals with heavy canvas

    b. by an NFPA-certified manufacturer

    c. by a relative, if they have a heavy-duty sewing machine

    d. by nobody

53. "Black fire" is apparent when a team of firefighters prepares to initiate fire attack. The smoke is high volume with a turbulent velocity and is ultradense and black. What should the firefighters expect to occur when they "read" this smoke condition?

    a. Backdraft

    b. Flash point

    c. Autoignition and flashover

    d. The decay phase

54. You are responding to an MVA with trapped occupants and extrication required. It is a hot July day. It is 1 p.m. and you have just returned from fighting a large dumpster fire. The temperature outside is 100°F and it is getting hotter. Which is the appropriate course of action concerning PPE?

    a. Wait until arrival on scene, then ascertain the proper amount of PPE to don, leaving off what is deemed unnecessary.

    b. Don full PPE even though your liners are a little moist. Leave jacket and trousers unfastened to allow ventilation.

    c. Pull the liners out of your PPE and don the outer shell and SCBA.

    d. Don full PPE and drink water en route to the scene. Dress down on scene, if allowed and appropriate.

55. You are preparing to hoist a chainsaw to the roof of a three-story building. Which of these knots is acceptable for this operation?

    a. clove hitch

    b. Becket bend

    c. figure-eight follow-through

    d. rescue knot

56. You and your two partners are advancing a charged 1¾" line to the second floor fire. At the bottom of the stairs, you see fire at the top of the stairs. You should _____.

    a. advance the hose to the top of the stairs and apply water

    b. apply water from the bottom of the stairs

    c. advance the hose to the midpoint of the stairs and apply water

    d. retreat and find another avenue to combat the fire

57. One of your fellow firefighters has been on vacation and returns to work with a substantial amount of facial hair. According to NFPA 1404 and _____, he should not be allowed to don and use an SCBA until the facial hair is removed.

    a. SCBA rules

    b. SCBA manufacturers' recommendations

    c. your fire marshal

    d. you

58. You respond to a possible residential structure fire call and find that a teenager has burned popcorn in the microwave oven. There is no fire, but the smell of burnt popcorn smoke is throughout the first floor of the house. As a customer service, you should consider _____.

    a. taking the microwave to the dumpster

    b. PPV of the unaffected portions

    c. PPV of the affected portions

    d. removing all popcorn so the episode cannot be repeated

59. You are on scene of a residential fire in a multistory building. The fire was on the fifth floor, has been knocked down, and you are checking for fire extension on the sixth floor. Command is delivering a lock puller tool to you. The only tools you have are a flat head ax and a flashlight. Your partner also has a pair of locking-type pliers and a screwdriver in his bunker coat pocket. You find you need to enter a locked door on the apartment directly above the apartment that was burning. You should _____.

    a. wait for the K tool

    b. try to force the door with the axe

    c. try wrenching the lock

    d. drive the screwdriver through the door with the axe

60. You are tasked with fire attack of a residential room and contents fire. It is a snowy wintry night and the temperature is hovering around 28°F. The ventilation crews confirm that a roof hole close to the seat of the fire is complete. You see that there is not much relief from the heat and smoke. You should consider _____.

    a. a retreat

    b. a direct attack

    c. an indirect attack

    d. hydraulic ventilation

61. A team of firefighters are placing an extension ladder to make entry into the structure and assist in the search of the second story of a residential structure. How will they place the tip of the ladder at the window?

    a. It should be placed one foot to the left or right of the window.

    b. It should be placed with the top rung inside the window frame.

    c. It should be placed with the top two rungs in the window frame.

    d. It should be placed level or no more than a few inches below the windowsill.

62. Heat comes from four basic sources. Which of the following is an example of a chemical heat source?

    a. Wooden materials on fire

    b. Friction

    c. Induction heating

    d. Nuclear power

63. You are fighting a grass fire that is downhill from the position of the engine. You know that you will _____ head pressure in this operation.

    a. gain

    b. lose

    c. modify

    d. not need

64. You are driving a rescue truck to the scene of a motor vehicle accident on a busy highway. Your first priority, upon arrival, should be to _____.

    a. position forward of the scene 100 feet

    b. position behind the scene 100 feet

    c. access the scene and establish a work area

    d. position so as not to impede traffic flow

65. When using a portable fire extinguisher, you know it is a good idea to _____.

    a. use the largest extinguisher available

    b. start fighting the smoke first

    c. quickly locate a second extinguisher

    d. look for an exit way behind you

66. Concerning the duty uniform, it is important to remember the duty uniform

    a. is an important component in protection from IDLH atmospheres

    b. can protect the wearer from IDLH atmospheres in the event that proper PPE cannot be donned

    c. adds protection for structural and proximity firefighting, but is unacceptable additional protection under the wildland PPE ensemble

    d. offers no protection for IDLH atmospheres under any circumstances

67. As a conscientious firefighter, you realize that dynamic influences can reduce the TPP of your structural gear. Fabric compression is one of these factors resulting from _____.

    a. folding your structural pants and placing them under your helmet while the gear is in storage

    b. wearing webbing that carries an axe and a flashlight

    c. wearing an SCBA

    d. all of the above

68. Pressurized Class B and Class C fires are similar in firefighting technique as you know you must _____.

    a. use PASS

    b. sweep the base of the fire

    c. shut off the source

    d. use a non-conducting nozzle

69. You have responded to a motor vehicle accident: car vs. car. It is 12 noon and the temperature is 100°F. The accident has occurred on a busy interstate highway. You and your two partners stabilize the cars and find that gas is leaking from one of the gas tanks and a rapid rate. The vehicle is a large SUV so you assume that there is potential of 30 or more gallons to leak out. You should be most concerned with

_____.

a. patient care

b. suppressing vapors

c. leak stoppage

d. airbag deployment

70. You are on a public service call to a residence in your fire district. The owner is concerned that her smoke detector will not operate when the power is out. You explain to her that she should consider a _____ detector.

a. battery-operated

b. hard-wired

c. combination

d. CO

71. _____ is an important concept for firefighters because it affects how to enter and function in a room or area that is on fire or where the fire has just been extinguished: firefighters must stay low to the floor in attacking structural fires.

a. Backdraft

b. Rollover

c. Thermal layering

d. Thermal conductivity

72. When you are using a BC fire extinguisher, you know _____.

a. the agent and nozzle will not conduct electricity

b. the agent will not conduct electricity, but the nozzle will

c. the nozzle will not conduct electricity, but the agent will

d. you are using the wrong extinguisher

73. You are securing your life-safety rope around an anchor and are preparing to rappel down a 10-foot drop off. There are no trees or other suitable anchors in the area. You decide to use the axle of your apparatus. You use which knot to accomplish this anchor?

a. figure-eight follow-through

b. Becket bend

c. clove hitch

d. half hitch

74. Two firefighters are deploying a large treated-canvas salvage cover on a group of items they have placed in the center of the room. Which of the following statements is false concerning this operation?

    a.   The treated side of the salvage cover should be on the outside when deployed.

    b.   Once the group of items is covered, the salvage cover edges should be rolled out.

    c.   Extra care should be taken if there are breakables in the stack of items being covered.

    d.   Covering the items in the room does not just apply to furnishings; it can also apply to carpeting.

75. Which of the following is a strategy used in salvage operations?

    a.   Moving an item out of harm's way

    b.   Covering an item to protect it

    c.   Removing the harmful substances from an area to protect other materials

    d.   All of the above

76. You have been charged with search and rescue of a doctor's office with a fire alarm activation but no smoke or flames showing. You have responded to this location numerous times for fire alarms, and they have all proved to be false. You and your partner prepare to enter the structure. You arrive at the entry door, and see through the window there is no fire or smoke, and no evidence of any damage. You do not see any occupants. Your first action is to check to see if _____.

    a.   there are auxiliary locking devices

    b.   there is a handle or knob on the outside of the door

    c.   there are exposed hinges

    d.   your partner has the K tool

77. The oxygen concentration in a closed compartment is 12 percent. Given this condition, what is true about combustion in the compartment.

    a.   There is not enough oxygen present to support combustion.

    b.   There is just enough oxygen present to support combustion.

    c.   The oxygen concentration is too high to support combustion.

    d.   The amount of oxygen is not related to combustion.

78. You are working a multifamily, multistory structure fire that is in defensive mode. You are assigned the task of putting water in an eighth story window. You know that you will need to deploy _____.

    a.   a 2½" line and a straight ladder

    b.   a master stream

    c.   a 1¾" line and a solid tip nozzle

    d.   an LDH attack line

79. You grab a section of 1¾" hose that is rolled and stored in your engine. You are going to connect it to one of the discharges of the engine. Usually this type of hose is rolled with the _____ end inside.

    a. male

    b. female

    c. other

    d. outside

80. A firefighter is checking a wall inside a structure for signs of hidden fire. Which of the following would indicate a hidden fire in the wall?

    a. The wall feels hot to the touch.

    b. A burning smell is noted near the outlets on the wall.

    c. The sound of items burning can be heard when near the wall.

    d. All of the above are signs of hidden fire in the wall.

81. You are operating an engine with two hoselines in use. Your officer asks you to charge another line for exposure protection. You will know if this is or is not possible because you marked the _____ pressure before charging any lines.

    a. residual

    b. static

    c. flow

    d. none of the above

82. During the daily check of your SCBA on your engine, you notice that the cylinder and regulator gauges differ. The cylinder gauge reads full, but the regulator gauge reads 1/2. You should _____.

    a. immediately fill the bottle until the regulator gauge reads full

    b. tag the SCBA, but as it is your unit, you should go ahead and use it for the shift, as you know the bottle is full

    c. replace the bottle and check the status of the gauges

    d. remove the pack from service

83. You respond to a reported smoke detector activation in the diorama room of a museum. Upon arrival, you are told there is a sprinkler system, but it has not activated. Based on this information, you can be relatively sure that the system is a _____ system.

    a. deluge

    b. dry pipe

    c. wet pipe

    d. pre-action

84. Several firefighters are discussing the growth stage of fire. The speed of the growth and ultimately the size of the fire are dependent on which of the following factors?

    a. Oxygen supply

    b. Fuel

    c. Container size

    d. All of the above

85. The ACGIH recommends that you only be exposed to 20 ppm of a chemical; while OSHA requires that the exposure be less than 35 ppm. To which one do you have to be legally adhered?

   a. OSHA

   b. ACGIH

   c. NIOSH

   d. None of the above

86. You are fighting a very large and hot room and contents fire in a residential structure.Because of poor visibility, your team decides to make a(n) _____.After this tactic, your team will probably have to back out of the fire area until the steam is dissipated.

   a. transitional attack

   b. indirect attack

   c. direct attack

   d. fire attack

87. Which of the following is a role of a public safety telecommunicator?

   a. Receiving emergency requests from citizens

   b. Evaluating the need for public safety response

   c. Alerting responders to the scene of emergencies

   d. All of the above

88. You and your partner are searching a commercial structure next door to a working structure fire. There is no smoke or fire in this building. You find a victim who appears to have fallen from a second story balcony. The appropriate rescue carry for this patient is _____.

   a. the firefighter's drag

   b. the webbing sling drag

   c. the clothing drag

   d. none of the above

89. You are venting the flat roof of a six-story apartment building and your chainsaw fails. You are unable to restart it. The attack crews are staged and waiting for ventilation. Under direction of IC and, if possible, you should _____.

   a. advise attack crews to begin the assault on the fire; ventilation is not possible

   b. use a rope and a tool to ventilate the windows on the sixth floor

   c. use a knife or other tool to begin removing the roofing cover

   d. use a rope and a tool to ventilate the windows on the fire floor

90. You are using a fire extinguisher to fight a pile of wood that is on fire. You estimate there is a little more than one cubic foot of wood. Your extinguisher is a 2A10BC. You know you _____.

    a. will have to go for another extinguisher as the amount of agent in this extinguisher is inadequate

    b. have more than enough agent to combat this fire

    c. will have to go for another type of extinguisher to meet the ABC requirement

    d. will have to decon the area after the fire is out as you are using dry chemicals considered to be hazardous

91. You respond to a reported smoke detector activation in the mainframe computer room of a school. Upon arrival, you are told there is a sprinkler system, but it has not activated. Based on this information, you can be relatively sure that the system is a _____ system.

    a. deluge

    b. dry pipe

    c. wet pipe

    d. pre-action

92. You are carrying the standpipe pack up the training tower. It feels lighter than regular hose to you. The reason is _____.

    a. you have become stronger

    b. it is single-jacket hose

    c. it is a polyester-blend hose

    d. it is much shorter hose

93. Which extremely toxic biological material is the easiest to make?

    a. Anthrax

    b. Ricin

    c. Tularemia

    d. RDD

94. You are responding to a wildland incident. While donning your boots, you break the shoelaces on one boot. What would your appropriate course of action be?

    a. Pull the lace out and tie it around the top of the boot to keep the boot on your leg.Request a lace from the quartermaster that evening.

    b. Borrow an extra lace or heavy string. Lace it through the eyelets and continue donning equipment. Replace both laces as soon as possible.

    c. Leave the boot loose, quickly don all PPE, and start hiking into the fire. Ask for shoelaces from coworkers on the way into the fire.

    d. Wrap heavy tape around your foot and ankle to keep the boot in place. Hike into the fire. Request a lace from the quartermaster that evening.

95. The chief of the Delmar Fire Department wants to completely change the way the department staffs the apparatus and stations. What will most likely be affected by the changes the chief would like to implement?

    a. Regulations

    b. Policies

    c. Bylaws

    d. Procedures

96. The only information that is available to you is a placard. Which book would be helpful?

    a. DOT ERG

    b. NIOSH

    c. CHRIS

    d. SBIMAP

97. You are on a rescue operation of a single-family residence with an unconscious person inside. There are security bars and gates on all windows and doors. There has been a flash flood and the home now has about one foot of water in it. Because of the layout and the terrain, the garage is still dry, but the water is rising. You have just used a rotary saw to cut into a garage door. You find that the door into the home from the garage is secured with a padlock on the garage side of the door. Your partner is carrying a flashlight and has no other tools. You should _____.

    a. use the saw to cut the padlock, one shackle at a time

    b. use the saw to cut the padlock, both shackles at the same time

    c. radio command for an alternate means to cut the lock

    d. return to staging

98. You ask your rookie firefighter to send you a pike pole for a training exercise in the training tower. You see her put a figure-eight knot in the rope. You remind her that the best knot for this operation is a _____.

    a. Becket bend

    b. figure-eight follow-through

    c. clove hitch

    d. bowline

99. You are on the scene of a multifamily residential structure fire in the middle of the night. A search team has found no occupants on the fire floor. Using the acronym REVAS, you should now begin _____ operation

    a. rescue

    b. evacuation

    c. ventilation

    d. exposure

100. You are filling an SCBA bottle from a truck-mounted Cascade unit. Your bottle is almost completely empty. You should open the valve of the cascade cylinder _____.

   a.   with the lowest pressure

   b.   quickly if the Cascade unit is equipped with a bottle cooler system

   c.   with the highest pressure

   d.   only if you are sure the SCBA bottle valve is closed

Phase III, Exam I: Answers to Questions

1. F	26. F	51. A	76. C
2. F	27. T	52. A	77. A
3. F	28. F	53. C	78. B
4. F	29. C	54. D	79. A
5. T	30. B	55. C	80. D
6. T	31. D	56. B	81. B
7. F	32. B	57. B	82. C
8. F	33. B	58. C	83. D
9. F	34. C	59. C	84. D
10. F	35. D	60. D	85. A
11. T	36. A	61. D	86. B
12. T	37. A	62. A	87. D
13. F	38. D	63. A	88. D
14. T	39. D	64. C	89. D
15. F	40. D	65. D	90. B
16. T	41. B	66. A	91. D
17. F	42. B	67. D	92. B
18. T	43. D	68. C	93. B
19. F	44. C	69. B	94. B
20. F	45. A	70. C	95. B
21. F	46. B	71. C	96. A
22. T	47. B	72. A	97. C
23. T	48. D	73. A	98. C
24. F	49. D	74. B	99. D
25. F	50. A	75. D	100. A

Phase III, Exam I:
Rationale & References for Questions

Question #1. Class I systems are designed for use by the fire department or trained personnel such as a fire brigade...No hose is provided for this class. NFPA 1001: 6.5.1. *FFHB, 2E:* Page 329.

Question #2. Any action designed to break the accident chain is known as an intervention. Intervention is typically a reactive action. Mitigation is a proactive action. NFPA 1001: 6.1.1.1. *FFHB, 2E:* Page 110.

Question #3. Almost anything in the path of the fire can be called an exposure. NFPA 1001: 5.3.8. *FFHB, 2E:* Page 609.

Question #4. In this situation, when applying the concept of risk/benefit, you can assume that the firefighters will not take significant risks to their safety to save what is already lost. This is especially true in light of the reports that there are no victims inside and very little property to save. NFPA 1001: 5.3; 6.3. *FFHB, 2E:* Page 730.

Question #5. In selecting the correct ladder for a structure, firefighters can use estimates of normal heights of floors for residential and commercial structures. Residences measure approximately 8-10 feet from floor to floor. The average commercial story will average approximately 10-12 feet from floor to floor. NFPA 1001: 5.3.6; 5.3.9; 5.3.10; 5.3.11. *FFHB, 2E:* Page 384.

Question #6. There should always be two easily recognizable ways off a roof. NFPA 1001: 5.3.11. *FFHB, 2E:* Page 581.

Question #7. Dry chemicals are very effective due to their coating action that reduces the chances of reignition. The coating action is a drawback, however, when protecting sensitive items such as computer mainframes. NFPA 1001: 5.3.16. *FFHB, 2E:* Page 188.

Question #8. They should also be inspected tactilely by running the rope through your hands. NFPA 1001: 5.1.1.1. *FFHB, 2E:* Page 440.

Question #9. Over-the-head method; grasp the back plate with both hands. Lift the back plate/cylinder overhead. NFPA 1001: 5.3.1. *FFHB, 2E:* Page 154.

Question #10. Connecting a hard sleeve to a hydrant can also be done, but it is not recommended due to the chance of drawing a vacuum on the hydrant or piping and causing damage.The hard sleeve is removed from its bed and attached to the inlet of the engine. The engine is then moved forward toward the hydrant with several firefighters carrying and positioning the sleeve.When the hard sleeve is aligned with the outlet, the engine is stopped and the sleeve connected. NFPA 1001: 5.3. *FFHB, 2E:* Page 264.

Question #11. Ventilation with this combination attack controls the flow of fire gases and steam, improving the survival chances of victims, and makes this attack the type typically used by firefighters in structural firefighting. NFPA 1001: 5.3 NFPA 6.3. *FFHB, 2E:* Page 290.

Question #12. A fire inspector from the Delmar Fire Department is conducting a fire prevention inspection in a new business. When he arrives, the business owner does not allow him to enter the building or conduct his inspection as planned. He will have to obtain an administrative warrant to enter the building without the consent of the business owner. NFPA 1001: 5.5.1. *FFHB, 2E:* Page 662.

Question #13. NFPA472: 5.3.2 - 5.3.4. *FFHB, 2E:* Page 897.

Question #14. Called a window search, this primary search takes advantage of speed by opening windows of rooms uninvolved and doing a quick look into the room. NFPA 1001: 5.3.9. *FFHB, 2E:* Page 465.

Question #15. Understanding various protective actions is important for responder health and safety. NFPA 472: 5.2.2 - 5.2.4. *FFHB, 2E:* Page 855.

Question #16. As a building burns, the structural elements decompose and lose their strength. This causes a change in the forces and the way the design loads are applied, leading to structural failure and collapse. NFPA 1001: 5.3.10; 6.3.2. *FFHB, 2E:* Page 345.

Question #17. The inspection of fire extinguishers by firefighters is usually a visual inspection. If something does not look right, the unit should be removed and replaced. NFPA 1001: 5.3.16. *FFHB, 2E:* Page 198.

Question #18. Damage from improper actions by firefighters include opening and closing valves or hydrants too quickly (thus creating a water hammer), failing to open or close a hydrant or other valve fully, or cross-threading the threads on the connections. NFPA 1001: 5.3.15. *FFHB, 2E:* Page 216.

Question #19. Although the incident commander maintains the responsibility for safe control of an incident, he or she may delegate authority for a function or task, but cannot transfer the responsibility for safe control to a subordinate. NFPA 1001: 6.1.1.1. *FFHB, 2E:* Page 38.

Question #20. Surface-to-mass ratio is a variable that can determine the amount of time a material can resist gravity and fire degradation. A material with a large surface area and small mass will be quicker to burn or fail. NFPA 1001: 5.3.10; 6.3.2. *FFHB, 2E:* Page 346.

Question #21. The American fire helmet was developed by a leather hat maker named Andrew Gratacap. The original hats were shaped like top hats and had the name of the fire company painted on them. Later, the idea of a short front brim and an extended back brim was incorporated into the design to assist in water runoff. NFPA 1001: N/A. *FFHB, 2E:* Page 11.

Question #22. Common practice is for the first-in unit to go to the control room. This room contains items that will assist in the determination of the location and severity of the fire. NFPA 1001: 5.5. *FFHB, 2E:* Page 616.

Question #23. By-products of combustion, such as toxic chemicals, particulate matter, and heat, can cause damage to and diseases of the lung tissue. The use of protective breathing apparatus in fires is required because of this danger. NFPA 1001: 5.3; 6.3. *FFHB, 2E:* Page 83.

Question #24. Atoms lacking electrons will tend to be quick to link up and form molecules. Atoms that are stable will tend to be those that have a satisfactory balance between the electron/proton ratio. NFPA 1001: 5.3; 6.3. *FFHB, 2E:* Page 80.

Question #25. The "removing items from harm's way" approach can be determined quickly by asking one simple question: Can it be moved quicker than it can be covered? If the answer is yes, firefighters should move the items. Items that are in the way of the fire attack crews should be moved to assist the operation and also accomplish salvage. NFPA 1001: 5.3.14. *FFHB, 2E:* Page 642.

Question #26. Appropriate BSI precautions must be taken for all patients. NFPA 1001: 4.3. *FFHB, 2E:* Page 698.

Question #27. A dry hydrant is not really a fire hydrant, but a connection point for drafting from a static water source such as a pond or stream, not a pressurized one. They are used primarily in rural areas with no water system. They may be found in urban and suburban areas as a backup or where certain buildings may be of great distance from other water sources, but a pond or stream is located nearby. NFPA 1001: 5.3.15. *FFHB, 2E:* Page 208.

Question #28. Telecommunicators must be able to prioritize incoming lines in order to ensure that the most important call gets the fastest attention. Incoming telephone calls should be answered in the following priority: 1. emergency or 9-1-1 lines 2. direct lines 3. business or administrative lines NFPA 1001: 5.2.1; 5.2.2. *FFHB, 2E:* Page 51.

Question #29. Firefighters arrive on a scene to find a Class C fire. The best method of attack is to remove the flow of electricity and then re-analyze the class of the fire for further actions. A Class A or B fire may be left after the Class C fire is resolved. NFPA 1001: 5.3; 6.3. *FFHB, 2E:* Page 100-101.

Question #30. NFPA requires manufacturers to clearly label care instructions for cleaning each piece of equipment. In addition, manufacturers should provide the user with specific instructions and information that addresses the limitations of use. NFPA 1001: 5.1.1.2. *FFHB, 2E:* Page 136.

Question #31. Open-circuit supplied air respirators (SARs), also called airline respirators, are similar to SCBA units, except that the air supply cylinder is remote from the user. Air is supplied in the same manner as for a regular SCBA, but the hose connecting the cylinder and the SCBA unit may be 100 to 200 feet long. This type of unit must be equipped with an SCBA escape unit with duration of approximately 5 to 10 minutes. NFPA 1001: 5.3.1. *FFHB, 2E:* Page 158.

Question #32. Understanding of basic chemical and physical properties is important for the health and safety of emergency responders. NFPA 472: 5.2.2 - 5.2.4. *FFHB, 2E:* Page 800.

Question #33. Hydrocarbons cover a wide range of substances in forms from gaseous to liquid to semisolid and solid. Common examples are heating oil, diesel fuel, gasoline, kerosene, petroleum jelly, paraffin, and asphalt. NFPA 1001: 5.3 NFPA 6.3. *FFHB, 2E:* Page 297.

Question #34. The danger inherent in a nuclear fuel fire results from exposure to radiation. NFPA 1001: 5.3; 6.3. *FFHB, 2E:* Page 85.

Question #35. The running end is used for work such as hoisting a tool. NFPA 1001: 5.1.1. *FFHB, 2E:* Page 425.

Question #36. PPE provides a minimum level of protection and should be considered the last resort of protection for firefighters and emergency responders operating at an incident. NFPA 1001: 5.1.1.2. *FFHB, 2E:* Page 125.

Question #37. As security requirements increase, home and business owners have begun to install many varieties of locks and security devices. If these types of devices are in use, the forcible entry team may need to find an alternate means of entry or use a rotary saw to gain access. NFPA 1001: 5.3.4; 6.3.2. *FFHB, 2E:* Page 530.

Question #38. Pressurized flammable liquids and gases are special fire hazards that should not be extinguished unless the fuel can be immediately shut off. NFPA 1001: 5.3.16. *FFHB, 2E:* Page 186.

Question #39. Common overhaul tools are pike poles, pitchforks, rubbish hooks, shovels, axes, chain saws, carry-alls, and wheelbarrows. NFPA 1001: 5.3.13. *FFHB, 2E:* Page 649.

Question #40. Other natural sources of surface water are rivers, lakes, and ponds. Drafting The pumping of water from a static source by taking advantage of atmospheric pressure to force water from the source into the pump NFPA 1001: 5.3.15. *FFHB, 2E:* Page 204.

Question #41. This system allows the pump to operate at full flow, but may cause damage to the pump and valves without proper back-flushing of the entire pump. NFPA 1001: 5.3 NFPA 6.3. *FFHB, 2E:* Page 298.

Question #42. Check for proper seal by attaching the regulator and inhaling a breath, activating the regulator.Hold that breath for about five seconds and listen and feel for any air leaks. NFPA 1001: 5.3.1. *FFHB, 2E:* Page 166.

Question #43. When folding salvage covers, firefighters should consider the deployment method that will be used for the covers, the amount of space for storing the covers on the apparatus, and folds that will be neat and orderly. NFPA 1001: 5.3.14. *FFHB, 2E:* Page 637-638.

Question #44. It is generally of little consequence in what order the occupancy is inspected as long as the method is efficient, systematic and thorough. Both firefighters can do the inspection in the order they want, as long as they cover everything and are systematic about their duties. This, of course, is subject to local standard operating procedures. A department can mandate that inspections be done in a certain pattern, in which case, that is the way all inspections must be performed. NFPA 1001: 5.5.1. *FFHB, 2E:* Page 662.

Question #45. In addition to the aluminized fabric, proximity PPE features full face shields that are coated with an anodized gold material to help create a "mirrored" reflective surface. Without this special coating, the wearer could receive radiant facial burns, and the face shield could quite possibly melt. NFPA 1001: 5.1.1.2. *FFHB, 2E:* Page 130.

Question #46. The SCBA unit and protective equipment add weight and bulk to the firefighter, causing increased exertion with loss of body fluids through perspiration. These actions increase during firefighting operations. Firefighters must be aware of them and of the symptoms of heat stress and their own limitations and abilities. During rehabilitation, EMS personnel should monitor vital signs and firefighters must hydrate to replace body fluids. NFPA 1001: 5.3.1. *FFHB, 2E:* Page 168, 169.

Question #47. Firefighters should understand their risks from hazardous materials exposures. NFPA 472: 4.2.1. *FFHB, 2E:* Page 743.

Question #48. Dump sites should be selected for availability to unload multiple tenders, turnaround area for the tenders, operational area, continued access to the fireground, and safety of personnel. NFPA 1001: 5.3.15. *FFHB, 2E:* Page 211.

Question #49. Firefighters need to create a fitness and wellness lifestyle that includes: cardiovascular conditioning, core strengthening, flexibility improvement, resistance training and nutritional balance. NFPA 1001: 5.3; 6.3. *FFHB, 2E:* Page 727.

Question #50. Most materials found in the passenger compartment are plastic, a form of polyvinyl chloride (PVC). PVC is extremely toxic when burned. NFPA 1001: 5.3.7. *FFHB, 2E:* Page 600.

Question #51. Firefighters arrive on a scene to find a Class A fire. The best method of attack is to cool the fire with water. NFPA 1001: 5.3; 6.3. *FFHB, 2E:* Page 100.

Question #52. This customizing should only be performed by a certified manufacturer that understands and complies with NFPA standards for both structural PPE and life safety system components. NFPA 1001: 5.1.1.2. *FFHB, 2E:* Page 128.

Question #53. "Black fire" is apparent when a team of firefighters prepares to initiate fire attack. The smoke is high volume with a turbulent velocity and is ultradense and black. They should expect autoignition or flashover when they "read" this smoke condition. NFPA 1001: 5.3; 6.3. *FFHB, 2E:* Page 103.

Question #54. Safety Firefighters should don all PPE necessary for the potential worst-case scenario. Granted, this approach may lead to "overdressing" for an incident. In these cases, the firefighters' company officer, incident commander, or incident safety officer may allow firefighters to "dress-down." NFPA 1001: 5.1.1.2. *FFHB, 2E:* Page 137.

Question #55. In general, anything that has a closed handle can be hoisted with a figure-eight or bowline, while longer cylindrical tools (i.e., ax and pike pole) can be hoisted using a clove hitch and half hitches. NFPA 1001: 5.1.1.1. *FFHB, 2E:* Page 449.

Question #56. The line is brought to the entrance or landing as described earlier, charged, and advanced at the point where the attack is to be made. Care is taken to ensure that sufficient hose and personnel are available to move to the next floor without having to stop on the stairs. The nozzle is operated to darken down the fire at the next landing, then shut down, and quickly advanced to that level, where it is opened again to secure the landing and moved forward to maintain the floor. NFPA 1001: 5.3. *FFHB, 2E:* Page 255.

Question #57. In addition, 29 CFR 1910.134, NFPA 1404 and SCBA manufacturers' recommendations prohibit any facial hair, which may interfere with proper fit and seal of the face piece. NFPA 1001: 5.3.1. *FFHB, 2E:* Page 166.

Question #58. The positive-pressure technique actually injects air into the compartment and pressurizes it. In an attempt to equalize, the smoke and heat are carried out into the areas of lower pressure outside the structure. NFPA 1001: 5.3.11. *FFHB, 2E:* Page 573.

Question #59. This method is not as quick as pulling the cylinder with a lock puller and should be used only if time allows. NFPA 1001: 5.3.4; 6.3.2. *FFHB, 2E:* Page 536.

Question #60. Great success can be achieved with positive pressure ventilation or with the use of a hoseline stream to create negative pressure and pull the smoke out. NFPA 1001: 5.3.11. *FFHB, 2E:* Page 588.

Question #61. If the ladder is used for access or escape, the tip should not extend into the window frame. The ideal location for access is for the ladder tip to be level or slightly below (no more than a few inches) the windowsill. NFPA 1001: 5.3.6; 5.3.9; 5.3.10; 5.3.11. *FFHB, 2E:* Page 385.

Question #62. The most common of the four sources of heat that firefighters deal with on a regular basis is the chemical reaction that releases heat as a by-product. Anything that burns does so through a chemical reaction in which heat is released as the bonds of the molecules break down. NFPA 1001: 5.3; 6.3. *FFHB, 2E:* Page 82-85.

Question #63. Head pressure measures the pressure at the bottom of a column of water in feet. Head pressure can be gained or lost when water is being pumped above or below the level of the pump. NFPA 1001: 5.3 NFPA 6.3. *FFHB, 2E:* Page 291.

Question #64. Based on the scene assessment, the first-arriving apparatus should be positioned to create a traffic barrier to help shield the greatest number of rescuers. NFPA 1001: 6.4.1. *FFHB, 2E:* Page 487.

Question #65. Firefighters must remember to have a path of escape behind them and not allow the fire to get between them and the exit. NFPA 1001: 5.3.16. *FFHB, 2E:* Page 195.

Question #66. A work uniform that meets NFPA standards is not designed to protect the wearer from IDLH atmospheres, but can add another layer of reasonable protection under wildland, proximity, and structural ensembles. NFPA 1001: 5.1.1.2. *FFHB, 2E:* Page 136.

Question #67. Dynamic influences such as dirty gear, moisture (including perspiration), and fabric compression (from an SCBA) can reduce the TPP of the clothing. NFPA 1001: 5.1.1.2. *FFHB, 2E:* Page 128.

Question #68. Pressurized flammable liquids and gases are special fire hazards that should not be extinguished unless the fuel can be immediately shut off. The recommended method of fighting these fires is to turn off or disconnect the electrical power and then use an appropriate extinguisher, depending on the remaining fuel source. NFPA 1001: 5.3.16. *FFHB, 2E:* Page 186.

Question #69. Caution: The vapors given off by a flammable liquid can be that material's most hazardous profile, sometimes traveling great distances before finding an ignition source. NFPA 1001: 6.3.1. *FFHB, 2E:* Page 601.

Question #70. Smoke detectors are hard or permanently wired, battery operated, or a combination of hard wiring with a battery backup. NFPA 1001: 6.5.1. *FFHB, 2E:* Page 204.

Question #71. Thermal layering is an important concept for firefighters because it affects how to enter and function in a room or area that is on fire or where the fire has just been extinguished: firefighters must stay low to the floor in attacking structural fires. NFPA 1001: 5.3; 6.3. *FFHB, 2E:* Page 99.

Question #72. Class C extinguishers have extinguishing agents and hoses with nozzles that will not conduct electricity. NFPA 1001: 5.3.16. *FFHB, 2E:* Page 186.

Question #73. The follow-through figure-eight knot is very useful when attaching a utility or life-safety line rope to an object that does not have a free end available. NFPA 1001: 5.1.1.1. *FFHB, 2E:* Page 433.

Question #74. Salvage cover edges should be rolled under so that they won't collect any water or debris. NFPA 1001: 5.3.14. *FFHB, 2E:* Page 643.

Question #75. Salvage operations may entail moving the item out of harm's way, covering the item, or removing the harmful substance from the area. NFPA 1001: 5.3.14. *FFHB, 2E:* Page 639.

Question #76. The firefighter must determine which way the door swings. The common way of describing this is in relationship to the forcible entry team. Doors with exposed or visible hinges will swing toward the forcible entry team. NFPA 1001: 5.3.4; 6.3.2. *FFHB, 2E:* Page 523.

Question #77. The oxygen concentration in a closed compartment is 12 percent. There is not enough oxygen present to support combustion. Combustion can occur in oxygen concentrations of 14 percent or greater. NFPA 1001: 5.3; 6.3. *FFHB, 2E:* Page 86.

Question #78. Solid stream handlines can reach more than 70 feet and master streams about 100 feet. NFPA 1001: 5.3 NFPA 6.3. *FFHB, 2E:* Page 282.

Question #79. The straight or storage hose roll is the easiest with which to work. Start with the hose flat on the ground. From the male end, to protect the threads, roll straight to the opposite end. NFPA 1001: 5.1.1.1. *FFHB, 2E:* Page 231.

Question #80. Firefighters can look for obvious signs of hidden fire and use their senses by feeling the wall for heat, smelling around outlets and other openings for a scent of anything burning, and listening for sounds of items burning. NFPA 1001: 5.3.13. *FFHB, 2E:* Page 649.

Question #81. Prior to charging any lines, the static pressure is read on the main intake compound gauge.The pump operator then charges the first line with the desired volume, noting the pressure first.With this flow going, the operator again reads the intake gauge and gets the residual pressure or the remaining pressure left in the system after the flow and friction loss from the flow. NFPA 1001: 5.3.15. *FFHB, 2E:* Page 215.

Question #82. Compare the cylinder and regulator gauges. Gauge readings should be within 100 psi. NFPA 1001: 5.3.1. *FFHB, 2E:* Page 172.

Question #83. Pre-action systems are used in areas where the materials protected are of high value and water damage would be expensive, such as computer rooms and historical items. NFPA 1001: 6.5.1. *FFHB, 2E:* Page 321.

Question #84. Several firefighters are discussing the growth stage of fire. The speed of the growth and ultimately the size of the fire are dependent on oxygen supply, fuel, container size, and insulation. NFPA 1001: 5.3; 6.3. *FFHB, 2E:* Page 92.

Question #85. Understanding toxicology and health effects is important. NFPA 472: 5.2.2; 5.2.3. *FFHB, 2E:* Page 838.

Question #86. This method, if used indoors, will usually greatly disturb the thermal balance, cause a loss of visibility in the structure as heat and products of combustion are circulated downward. When using this method, the firefighting team should attempt to shield themselves from the steam vapor by backing out of the area after applying the water. They can reenter as soon as the vapor begins to dissipate. NFPA 1001: 5.3.10. *FFHB, 2E:* Page 604.

Question #87. The role of the public safety telecommunicator is to receive emergency requests from citizens, evaluate the need for public safety response, and alert responders to the scene of emergencies. NFPA 1001: 5.2.1. *FFHB, 2E:* Page 48.

Question #88. The rescuer must remember that none of these drags provides spinal immobilization and are intended to be utilized only in situations where greater harm will come to the patient if not immediately moved. NFPA 1001: 5.3.9. *FFHB, 2E:* Page 477.

Question #89. A technique that has produced satisfactory results when operating off a flat roof beyond the reach of a ladder is to use a tool and a rope. NFPA 1001: 5.3.11. *FFHB, 2E:* Page 570.

Question #90. For a 1-A rating, the extinguisher should extinguish a wood crib fire of about one cubic foot. The ratings increase as the amount of fire suppressed increases. For instance, a 2-A extinguisher will put out twice the fire of a 1-A. NFPA 1001: 5.3.16. *FFHB, 2E:* Page 194.

Question #91. Pre-action systems are used in areas where the materials protected are of high value and water damage would be expensive, such as computer rooms and historical items. NFPA 1001: 6.5.1. *FFHB, 2E:* Page 321.

Question #92. Standpipe packs often use single-jacket hose to reduce the weight of the pack. NFPA 1001: 5.3. *FFHB, 2E:* Page 222.

Question #93. First responders should have a basic awareness of the threat of terrorism and basic response actions. NFPA 472: 4.2.1. *FFHB, 2E:* Page 930.

Question #94. Additionally, a good fitting, tightly laced boot can help prevent ankle sprains and reduce foot fatigue. NFPA 1001: 5.1.1.2. *FFHB, 2E:* Page 132.

Question #95. Policies are formal statements or directives established by fire department managers to provide guidance for decision making. Staffing of apparatus and stations will likely be outlined in new department policies. NFPA 1001: 5.1.1.1. *FFHB, 2E:* Page 34.

Question #96. Specific chemical information is important for a safe and informed response. NFPA 472: 4.2.1 - 4.2.2. *FFHB, 2E:* Page 809.

Question #97. Attach locking-type pliers to the lock case and lock the jaws. The locking pliers must have a chain or rope attached so the firefighter can hold the pliers clear of the saw. A rotary saw with metal cutting blade is used to cut the lock shackles. Figure 17-50 (FFHB, 2E), Cut both sides of the shackle on one cut. NFPA 1001: 5.3.4; 6.3.2. *FFHB, 2E:* Pages 540-541.

Question #98. In general, anything that has a closed handle can be hoisted with a figure-eight or bowline, while longer cylindrical tools (i.e., ax and pike pole) can be hoisted using a clove hitch and half hitches. NFPA 1001: 5.1.1.1. *FFHB, 2E:* Page 449.

Question #99. Still another acronym is termed REVAS: Rescue, Exposures, Ventilation, Attack, and Salvage. NFPA 1001: 5.3.3. *FFHB, 2E:* Page 607.

Question #100. Open the valve of the cascade cylinder with the lowest pressure. This pressure must be higher than the pressure in the cylinder being filled. NFPA 1001: 5.3.1. *FFHB, 2E:* Page 178.

Phase Three, Exam Two

1. Combination headsets are considered adequate hearing protection.

   a. True

   b. False

2. First responders should always follow the "use water to knock down vapors" recommendation in the DOT ERG.

   a. True

   b. False

3. You are part of a search and rescue team on a commercial structure with a glass entry door. You decide breaking the glass will be the fastest way to gain entry to the structure. However, you know the glass is laminated and you will need find another entry point.

   a. True

   b. False

4. EHS materials do not present any risk to the community.

   a. True

   b. False

5. Oil and gasoline can be controlled by flushing the area with large amounts of water, performing dilution of the spill.

   a. True

   b. False

6. While enjoying an off-duty afternoon in a movie theater, you find a seat burning in the second row. You tell your significant other to call 9-1-1 while you search for an extinguisher. You find four extinguishers with the older symbols on them. You know you want to pick the extinguisher with a yellow star and a letter inside it.

   a. True

   b. False

7. A pressure tank that impinged by fire is not dangerous unless the relief valve is on fire.

   a. True

   b. False

8. The only protective clothing for which heat stress is an issue is Level A.

   a. True

   b. False

9. A group of firefighters is performing a fire safety inspection in a business. When they inspect the electrical hazards, they will check the fuse and breaker panels to verify that overcurrent protection devices have not been defeated and that all spaces on the panel are equipped with a breaker switch or blank to cover the opening.

    a. True

    b. False

10. An illegal drug lab is less dangerous than an illegal biological lab.

    a. True

    b. False

11. The type of fire, the characteristics of the burn, and the manner in which fuel is being supplied are the determining factors in deciding the best attack.

    a. True

    b. False

12. The Hazard Communication standard 29 CFR 1910.1200 requires that MSDS be maintained for chemical safety.

    a. True

    b. False

13. A fire inspector from the Delmar Fire Department is conducting a fire prevention inspection in a new business. When he arrives, the business owner does not allow him to enter the building or conduct his inspection as planned. He will have to obtain an administrative warrant to enter the building without the consent of the business owner.

    a. True

    b. False

14. You find an explosive device in a building. The best course of action is to take it outside so the building is not damaged.

    a. True

    b. False

15. Seat-mounted SCBA allow the unit to be donned en route to a fire.

    a. True

    b. False

16. You are tying off a life-safety rope for a rappelling operation. After tying your knot, you should set the knot to ensure it is tied correctly.

    a. True

    b. False

17. You are nozzle person fighting a basement fire. You cannot make a direct attack on the fire. Requesting a Bresnan distributor to assist with your attack operation would be a correct course of action.

    a. True

    b. False

18. While in the hot desert city of Phoenix, Arizona, one might assume that the city fire hydrants are dry barrel hydrants.

    a. True

    b. False

19. Out-of-service hose can be put back into service only if it is tested on its annual test due date.

    a. True

    b. False

20. You are directed by your IC to conduct a primary search. On your way to the entry door, you take a moment to look into an uninvolved room's window to check for occupants. This is considered part of the primary search.

    a. True

    b. False

21. When considering deployment methods for salvage covers, both the shoulder toss and balloon toss deployments are done by two firefighters.

    a. True

    b. False

22. In the rehabilitation area, passive cooling practices (shade, air movement, rest) are less effective in reducing heat stress in firefighters than active cooling activities, such as forearm immersion technique.

    a. True

    b. False

23. You are preparing to enter the hallway of an apartment complex that has had a sprinkler activation. You have requested a "stop" as you know this is the only way to shut down water flow from opened sprinklers.

    a. True

    b. False

24. Depth of char is a term commonly used by fire investigators to describe the amount of time wooden material has burned. The deeper the char, the longer the material was burning or exposed to direct flame.

    a. True

    b. False

25. You have responded to a commercial fire alarm in a local hardware/building supplies chain. The 2 A.M. alarm has used all available fire resources as there is another fire alarm in an apartment complex across town. The number one priority on both of these fire scenes is rescue/evacuation.

    a. True

    b. False

26. Trusses create large void spaces. The obvious collapse danger with these void spaces is that the fire may be undetected while it continues to compromise the structural elements.
    a. True
    b. False

27. Efficient use of time dictates the necessity of breaking windows rather than trying to open them.
    a. True
    b. False

28. You have responded to a multifamily, multistory residential structure fire in the downtown area of your district. The time is 2 A.M. and the weather is cold and snowy. The building is a care facility for the elderly. Upon arrival, you see smoke and flames coming from a fourth floor apartment. You see smoke coming from several of the other units on the same floor. The are occupants running from the building. A few of them have slipped and fallen on the icy sidewalk. Most are barefoot with no winter clothing and many are shivering. Most are exhausted from the rapid evacuation down the stairs. There are occupants on the balcony of the unit with flames showing and there are occupants on many of the balconies of the smoke-filled apartments. Your first priority would be to move the evacuated occupants to a warm environment.
    a. True
    b. False

29. You are in the stairwell of an apartment complex. There is smoke showing from the fourth floor. You are unsure where the fire is on the fourth floor, but know the first through third floors are clear. You decide to make the standpipe connection of your high rise pack on the _____ floor
    a. second
    b. third
    c. fourth
    d. fifth

30. Firefighter Mason is performing a personal size-up while she is donning her personal protective equipment in preparation to enter a burning structure. Which of the following would not be important in her personal size-up in this situation?
    a. An awareness of established work areas
    b. Active cooling
    c. Analysis of smoke conditions
    d. Identification of possible escape routes

31. You are venting the flat roof of a six-story apartment building and your chainsaw fails. You are unable to restart it. The attack crews are staged and waiting for ventilation. Under direction of IC and, if possible, you should _____.
    a. advise attack crews to begin the assault on the fire; ventilation is not possible
    b. use a rope and a tool to ventilate the windows on the sixth floor
    c. use a knife or other tool to begin removing the roofing cover
    d. use a rope and a tool to ventilate the windows on the fire floor

32. You are preparing to hoist a ladder to the third floor of a commercial building. You remember that you will use the _____ of the rope for work such as this.

    a. working end

    b. standing end

    c. standing part

    d. running end

33. You are on a public service call to a residence in your fire district. The owner is concerned that her smoke detector will not operate when the power is out. You explain to her that she should consider a _____ detector.

    a. battery-operated

    b. hard-wired

    c. combination

    d. CO

34. You are responding to a reported structure fire in a rural area outside of town. Your officer calls for water tenders to respond. She tells you the hydrants in this area have almost no pressure. She knows this from hydrant _____.

    a. size

    b. testing

    c. type

    d. shape

35. You ask your rookie firefighter to send you a pike pole for a training exercise in the training tower. You see her put a figure-eight knot in the rope. You remind her that the best knot for this operation is a _____.

    a. Becket bend

    b. figure-eight follow-through

    c. clove hitch

    d. bowline

36. You and your partner have exited a residential structure fire after conducting a primary search. You return to your engine to change out your bottles and you notice your partner is quite pale. What should you do?

    a. Tell her what you see and encourage her to drink some water or Gatorade.

    b. Tell her what you see and have her sit in the shade for a few minutes while drinking fluid replacement drinks.

    c. Tell her what you see and suggest she report to rehab.

    d. Assume she knows her ability. If she does not complain, then it is safe to assume she is fine and ready for another assignment.

37. You respond to a reported smoke detector activation in the diorama room of a museum. Upon arrival, you are told there is a sprinkler system, but it has not activated. Based on this information, you can be relatively sure that the system is a _____ system.
    a. deluge
    b. dry pipe
    c. wet pipe
    d. pre-action

38. You are carrying the standpipe pack up the training tower. It feels lighter than regular hose to you. The reason is _____.
    a. you have become stronger
    b. it is single-jacket hose
    c. it is a polyester-blend hose
    d. it is much shorter hose

39. You are explaining hydraulics to your rookie firefighters. You explain the formula for calculating engine pressure. One of the students does not understand the formula as you explain it. Therefore, you write it on the board as _____.
    a. $EP=NP-FL\pm E+SA$
    b. $EP=NP*FL\pm E+SA$
    c. $EP=NP/FL\pm E+SA$
    d. $EP=NP+FL\pm E+SA$

40. A group of preschool children have come to visit the fire station. Firefighter Williams is giving a tour of the facility. Which of the following things should be focused on during his presentation to the preschool children?
    a. How to call for the fire department in an emergency
    b. Fire department response procedures
    c. Proper placement of carbon monoxide detectors
    d. Upcoming first aid classes

41. You are operating an engine with two hoselines in use. Your officer asks you to charge another line for exposure protection. You will know if this is or is not possible because you marked the _____ pressure before charging any lines.
    a. residual
    b. static
    c. flow
    d. none of the above

42. Which of the following developments in the American fire service did not occur in the period during and after the Civil War?
    a. Dalmatians became known as mascots of the fire service.
    b. Firefighters banded together to form a labor union.
    c. The use of military rank structures and command and control tactics became common in the fire service.
    d. Horses began to be used to pull the heavy steam-powered fire apparatus.

43. You and your two partners are advancing a charged 1¾" line to the second floor fire. At the bottom of the stairs, you see fire at the top of the stairs. You should _____.

    a. advance the hose to the top of the stairs and apply water

    b. apply water from the bottom of the stairs

    c. advance the hose to the midpoint of the stairs and apply water

    d. retreat and find another avenue to combat the fire

44. You are flighting a wood pile fire at a cabinet factory. The fire is relatively small. You estimate it to be 1 cubic foot. The wind is not blowing, the temperature is 50°F, and the fire is not increasing in size rapidly. The nearest extinguisher is rated as a 30A160BC. There is a 2A 10BC extinguisher mounted on the wall 100 feet away. What should you do?

    a. Use the 30A160BC extinguisher to put out the fire.

    b. Wait for an attack line from the engine.

    c. Watch the fire while you send your partner for the 2A10BC extinguisher.

    d. Look for a Class A only extinguisher.

45. You are searching for the seat of a fire with your partner in the interior of a residential structure fire. As you advance down a hallway, you find you are following an exterior wall with windows and a door. The products of combustion are thick and visibility is just inches. However, you do not feel an increase in ambient temperature. Your course of action should be to _____.

    a. retreat and find other access to the fire

    b. notify IC and retreat to find other access to the fire

    c. break out windows as you advance

    d. hold your position at the mouth of the hallway and have your partner advance toward the fire

46. You are fighting a very large and hot room and contents fire in a residential structure.Because of poor visibility, your team decides to make a(n) _____.After this tactic, your team will probably have to back out of the fire area until the steam is dissipated.

    a. transitional attack

    b. indirect attack

    c. direct attack

    d. fire attack

47. Which of the following statements is true about the thermal conductivity of materials.

    a. A very dense material will conduct heat better than a less dense material.

    b. A very dense material will conduct heat the same as less dense material.

    c. A less dense material will conduct heat better than a very dense material.

    d. None of the above are correct.

48. One of your fellow firefighters has been on vacation and returns to work with a substantial amount of facial hair. According to NFPA 1404 and _____, he should not be allowed to don and use an SCBA until the facial hair is removed.

    a. SCBA rules

    b. SCBA manufacturers' recommendations

    c. your fire marshal

    d. you

49. An engine company is out conducting fire safety inspections in a row of businesses in their first due area. Firefighter Jackson is going to perform her inspection from left to right inside the building. Firefighter Wheatley says that he is going to perform his inspection from outside to inside, and then from top to bottom. Who is correct?

    a. Firefighter Jackson

    b. Firefighter Wheatley

    c. Both are correct.

    d. Neither are correct.

50. While fighting a wildland fire, you notice you and your crew have hiked onto a saddle above the advancing fire. It is 3 P.M. The weather is changing rapidly. Currently the wind is blowing the fire back down slope. You should _____.

    a. evacuate to a safe zone

    b. notify the plane of the weather change

    c. deploy your shelter

    d. establish a watch to monitor the weather

51. Which of the following parts of a ladder can be found on an aerial ladder apparatus but not on a ground ladder?

    a. Beams

    b. Rails

    c. Hydraulic pistons

    d. All of the above

52. You are on a rescue operation of a single-family residence with an unconscious person inside. There are security bars and gates on all windows and doors. There has been a flash flood and the home now has about one foot of water in it. Because of the layout and the terrain, the garage is still dry, but the water is rising. You have just used a rotary saw to cut into a garage door. You find that the door into the home from the garage is secured with a padlock on the garage side of the door. Your partner is carrying a flashlight and has no other tools. You should _____.

    a. use the saw to cut the padlock, one shackle at a time

    b. use the saw to cut the padlock, both shackles at the same time

    c. radio command for an alternate means to cut the lock

    d. return to staging

53. You are a firefighter assigned to the airport station. You respond to a reported in-flight emergency that might result in a fiery crash. En route to the staging point, you realize you have your structural helmet instead of your proximity helmet. Your concern would be _____.

    a. radiant facial burns from a melting shield

    b. your helmet is black and not silver

    c. less impact protection from special grade aircraft structural members

    d. "ARFF blindness" due to special coatings not being applied to your structural face shield

54. You have responded to a residential structure fire in a neighborhood that is known for its high crime rates. It is 2 A.M. and there is a car parked in front of the home on the street. The home has bars over the windows. You are on the search team and you attempt to force the front door for entry. You spend a minute or so attempting to force the door and it simply will not move. You _____.

    a. suspect an auxiliary locking device

    b. should keep trying to force the door

    c. should try pulling the lock

    d. should try using a K tool

55. Your department practice is to issue PASS devices to individual firefighters. You are donning PPE and realize you left your PASS device at the station.What should you do?

    a. Continue to don PPE and prepare for initial fire attack and say nothing about your missing PASS device.

    b. Realize that you will have to resort to the back-up PASS practice of screaming loudly if you get into trouble.

    c. Realize that not wearing a PASS device is as serious as not turning one on, and you should notify command immediately.

    d. Take a PASS device from another firefighter on scene and owe him or her a big favor.

56. A team of firefighters are placing an extension ladder to access a roof of a strip mall to check for fire extension and perform ventilation activities. How should they place the tip of the ladder?

    a. It should be placed six to twelve inches below the roof line.

    b. It should be placed even with the roof line.

    c. It should be placed one rung above the roof level.

    d. It should be placed above the roof level approximately five rungs.

57. You respond to a reported smoke detector activation in the diorama room of a museum. Upon arrival, you are told there is a sprinkler system, but it has not activated. Based on this information, you can be relatively sure that the system is a _____ system.

    a. deluge

    b. dry pipe

    c. wet pipe

    d. pre-action

58. You and your partner are searching a commercial structure next door to a working structure fire. There is no smoke or fire in this building. You find a victim who appears to have fallen from a second story balcony. The appropriate rescue carry for this patient is _____.

    a. the firefighter's drag

    b. the webbing sling drag

    c. the clothing drag

    d. none of the above

59. You have been charged with search and rescue of a doctor's office with a fire alarm activation but no smoke or flames showing. You have responded to this location numerous times for fire alarms, and they have all proved to be false. You and your partner prepare to enter the structure. You arrive at the entry door, and see through the window there is no fire or smoke, and no evidence of any damage. You do not see any occupants. Your first action is to check to see if _____.

    a. there are auxiliary locking devices

    b. there is a handle or knob on the outside of the door

    c. there are exposed hinges

    d. your partner has the K tool

60. How can a firefighter assist fire investigators in discovering how and why a fire occurred?

    a. Remove all of the contents of the burned rooms to the outside of the structure for ease in conducting investigations.

    b. Use large volumes of water to ensure materials are thoroughly soaked prior to fire investigation efforts.

    c. Being observant of the variety of conditions and occurrences on the fire scene and passing this information to fire investigators.

    d. All of the above are correct.

61. You are trying to put a swift water rescue throw bag together to keep in your bunker coat pocket. Your best choice, in this case, would be to use _____ rope.

    a. polyester

    b. polypropylene

    c. polyethylene

    d. nylon

62. A large warehouse fire has a total of 22 firefighters working in the Operations Section. Firefighter Smith says that it is acceptable to split the 22 firefighters into three divisions for an effective span of control. Firefighter Taylor says that the 22 firefighters need to be split into at least four divisions for an effective span of control. Who is correct?

    a. Firefighter Smith is correct.

    b. Firefighter Taylor is correct.

    c. Both are correct.

    d. Neither are correct.

63. The top of a beam is subjected to a(n) _____ force while the bottom of the beam is subjected to tension.

    a. axial

    b. imposition

    c. shear

    d. compressive

64. There are certain behavioral characteristics and traits that are needed to operate successfully as a public safety telecommunicator. Which of the following is not one of these characteristics or traits?

    a. The ability to perform one task at a time

    b. The ability to maintain composure in high-stress conditions

    c. The ability to remember details and recall information easily

    d. The ability to exercise voice control

65. Reports of emergencies can be received in a variety of ways. Which of the following is a way in which a report of emergency might be received?

    a. Emergency call boxes

    b. Wireless or cellular phones

    c. Automatic alarms

    d. All of the above

66. You are responding to a wildland incident. While donning your boots, you break the shoelaces on one boot. What would your appropriate course of action be?

    a. Pull the lace out and tie it around the top of the boot to keep the boot on your leg.Request a lace from the quartermaster that evening.

    b. Borrow an extra lace or heavy string. Lace it through the eyelets and continue donning equipment. Replace both laces as soon as possible.

    c. Leave the boot loose, quickly don all PPE, and start hiking into the fire. Ask for shoelaces from coworkers on the way into the fire.

    d. Wrap heavy tape around your foot and ankle to keep the boot in place. Hike into the fire. Request a lace from the quartermaster that evening.

67. Pressurized Class B and Class C fires are similar in firefighting technique as you know you must _____.

    a. use PASS

    b. sweep the base of the fire

    c. shut off the source

    d. use a non-conducting nozzle

68. You are working in a station that responds to wildland fires. You are folding the station laundry from the clothes dryer and you find your partner's wildland bandana in with the station bathroom towels. What should you do?

    a. Re-wash the towels to ensure no cross-contamination with products of combustion, and remind your partner of the hazards of this practice.

    b. Tell your partner that practice is not acceptable and in the future to keep his/her PPE separate from station laundry. Fold and put away the laundry.

    c. Dispose of the station bathroom towels.

    d. Put away the laundry and say nothing.

69. As a tender shuttle driver/operator, you should pick a dump site that has
    _____.

    a. access

    b. turnaround area

    c. safety

    d. all of the above

70. Your engine is attached to the hydrant using two 2½" supply lines. You notice that one of the lines is leaking a considerable amount of water at one of the couplings. You should _____.

    a. pull another section of 2½" line for replacement

    b. tighten the coupling with a hydrant wrench

    c. tighten the coupling with spanner wrenches

    d. use a hose bridge

71. You are a firefighter on scene of a car fire. There are flames showing from the grill and heavy smoke is billowing out from the wheel wells and from under the car. You watch two firefighters on a hoseline advance toward the engine compartment fire. They are advancing in a straight line directly in front of the car toward the grill. You should have them _____.

    a. wait until you open the hood

    b. approach from the side

    c. approach from the back

    d. wait until foam can be applied

72. You are responding to an MVA with trapped occupants and extrication required. It is a hot July day. It is 1 p.m. and you have just returned from fighting a large dumpster fire. The temperature outside is 100°F and it is getting hotter. Which is the appropriate course of action concerning PPE?

    a. Wait until arrival on scene, then ascertain the proper amount of PPE to don, leaving off what is deemed unnecessary.

    b. Don full PPE even though your liners are a little moist. Leave jacket and trousers unfastened to allow ventilation.

    c. Pull the liners out of your PPE and don the outer shell and SCBA.

    d. Don full PPE and drink water en route to the scene. Dress down on scene, if allowed and appropriate.

73. You are using a fire extinguisher to fight a pile of wood that is on fire. You estimate there is a little more than one cubic foot of wood. Your extinguisher is a 2A10BC. You know you _____.

    a. will have to go for another extinguisher as the amount of agent in this extinguisher is inadequate

    b. have more than enough agent to combat this fire

    c. will have to go for another type of extinguisher to meet the ABC requirement

    d. will have to decon the area after the fire is out as you are using dry chemicals considered to be hazardous

74. You are on a rescue scene. You are attempting to extend a rope for a rescue operation, as it is too short. You know that you have a duplicate rope in your apparatus. You know that you should consider the _____ to assist your operation.

    a. Becket bend

    b. bowline

    c. water knot

    d. figure-eight

75. Firefighter O'Malley is preparing for an upcoming presentation on fire safety to a high school class. Which of the following is not important for Firefighter O'Malley to consider in preparing for the presentation?

    a. The presentation must include current information.

    b. The presentation should try to relate the subject matterto the audience.

    c. The presentation should include some "hands-on" activities, if possible, to reinforce learned concepts.

    d. The presentation should end with a comprehensive test to check that the audience was paying attention.

76. On a residential structure fire scene in the winter, you notice that one of the attack lines has frozen. You should _____.

    a. heat it to thaw the water

    b. carefully move it to a place where it can thaw

    c. roll it and place it on the apparatus

    d. discard it

77. You are tasked with fire attack of a residential room and contents fire. It is a snowy wintry night and the temperature is hovering around 28°F. The ventilation crews confirm that a roof hole close to the seat of the fire is complete. You see that there is not much relief from the heat and smoke. You should consider _____.

    a. a retreat

    b. a direct attack

    c. an indirect attack

    d. hydraulic ventilation

78. Your incident commander has tasked you with roof ventilation for a defensive commercial structure fire operation. The building is a large, flat-roofed structure with several wings and command has given the order to confine the fire to the affected wing. Command is ready to write off the wing, but dictates that all companies will save the unaffected portions of the building. The roof covering is tar paper and the pilot hole reveals plywood sheeting. The roof is not spongy. In these conditions, you would direct your vent team to make _____.

    a. a large vent hole

    b. a louvered cut

    c. a triangle cut

    d. a trench cut

79. Which of the following parts of a ladder can be found on an extension ladder but not on a straight ladder?

    a. A halyard

    b. Dogs, pawls, rung locks, or ladder locks

    c. A pulley

    d. All of the above

80. You are fighting a grass fire that is downhill from the position of the engine. You know that you will _____ head pressure in this operation.

    a. gain

    b. lose

    c. modify

    d. not need

81. You are on scene of a residential fire in a multistory building. The fire was on the fifth floor, has been knocked down, and you are checking for fire extension on the sixth floor. Command is delivering a lock puller tool to you. The only tools you have are a flat head ax and a flashlight. Your partner also has a pair of locking-type pliers and a screwdriver in his bunker coat pocket. You find you need to enter a locked door on the apartment directly above the apartment that was burning. You should _____.

    a. wait for the K tool

    b. try to force the door with the axe

    c. try wrenching the lock

    d. drive the screwdriver through the door with the axe

82. You are driving a rescue truck to the scene of a motor vehicle accident on a busy highway. Your first priority, upon arrival, should be to _____.

    a. position forward of the scene 100 feet

    b. position behind the scene 100 feet

    c. access the scene and establish a work area

    d. position so as not to impede traffic flow

83. You are on scene of a car fire. You find the vehicle to be a diesel pickup truck, so you decide to use foam. You know that diesel is a _____ and this will determine the type of foam you will use.

    a. bio fuel
    b. hydrocarbon
    c. polar solvent
    d. Class A fuel

84. The lower flammable limit of kerosene is 0.7 and the upper explosive limit is 5. Which of the following readings for kerosene would be within the range where Kerosene could ignite?

    a. 0.5
    b. 3
    c. 7
    d. 10

85. You are filling an SCBA bottle from a truck-mounted Cascade unit. Your bottle is almost completely empty. You should open the valve of the cascade cylinder _____.

    a. with the lowest pressure
    b. quickly if the Cascade unit is equipped with a bottle cooler system
    c. with the highest pressure
    d. only if you are sure the SCBA bottle valve is closed

86. While doing an interior attack in a residential structure, you and your partner find it necessary to make your own escape from an interior bedroom. You decide your best course of action is to breech the wall and escape through an adjacent bedroom. You and your partner punch a hole in the drywall and find a safe egress. Your partner, who is smaller than you, has a bit of trouble fitting through the breech, and you decide you will have more difficulty. Your best course of action is to _____.

    a. wait for your partner to escape and go for help
    b. remove your SCBA and proceed through the breech while holding the SCBA in front of you
    c. attempt to breech another hole in a different location
    d. loosen the straps and rotate the SCBA under your arm and attempt to fit through the breech

87. A firefighter is performing an assessment on an unconscious adult woman who was discovered lying next to her home. The firefighter first made sure the scene was safe, then donned her BSI protective equipment. She performed a quick visual survey of the patient as she approached her, and then began her initial assessment. She checked for level of consciousness, airway, breathing, and circulation in the patient. What is the firefighter's next step?

    a. Assessing color and temperature of the skin
    b. Obtaining pulse and respiratory rates
    c. A focused history and physical exam
    d. Checking for major bleeding

88. While doing company inspections, the engine company finds a building with special egress control devices. What can they expect about the building?

   a.   The building will have an approved fire detection system.

   b.   This building will have an approved fire sprinkler system.

   c.   The special egress control devices will deactivate with a loss of electrical power in the building.

   d.   All of the above are correct.

89. Concerning the duty uniform, it is important to remember the duty uniform

   a.   is an important component in protection from IDLH atmospheres

   b.   can protect the wearer from IDLH atmospheres in the event that proper PPE cannot be donned

   c.   adds protection for structural and proximity firefighting, but is unacceptable additional protection under the wildland PPE ensemble

   d.   offers no protection for IDLH atmospheres under any circumstances

90. While doing an interior search of a commercial structure fire, your partner's low-air alarm activates. He notifies you that he needs to leave. You both retreat and follow your search pattern out of the building. About 200 feet from the exit, your partner notifies you he is out of air. You see him remove his regulator, pull his hood over the opening, and put his face to the floor. You should _____.

   a.   stop, notify IC that you are in trouble, and buddy breath until the "two out" team can rescue you

   b.   start buddy breathing and continue your emergency egress

   c.   maintain contact with your partner and continue your emergency egress

   d.   continue your emergency egress with the intent of sending in more air for him if you lost him on the way out of the building

91. You are in an IDLH atmosphere and your hear the order to evacuate on another firefighter's radio. You discover your radio is not operating. You have become disoriented and your partner has left you. You yell to gain anybody's attention and nobody responds._____ should alert IC to your absence.

   a.   Your partner

   b.   Your PASS device

   c.   The accountability system

   d.   All of the above

92. The only information that is available to you is a placard. Which book would be helpful?

   a.   DOT ERG

   b.   NIOSH

   c.   CHRIS

   d.   SBIMAP

93. What is a common choking agent that is found in most communities?

    a. Nerve agents

    b. Chlorine

    c. Sarin

    d. Anthrax

94. Firefighters Garcia and Watkins just returned from an incident where a child was injured in a severe automobile accident. Firefighter Garcia thinks that Firefighter Watkins is experiencing critical incident stress as a result of this last emergency incident. Which of the following could be an indication of critical incident stress?

    a. Firefighter Watkins is withdrawn for the rest of the shift.

    b. Firefighter Watkins is angry and hostile to his coworkers.

    c. Firefighter Watkins seems to care little about the job and to be fearful of the next incident.

    d. All of the above can be indications of critical incident stress.

95. An important agent maintenance duty for dry chemical extinguishers is to

_____.

    a. wash the exterior with soap and water

    b. visually examine the exterior for soap residue

    c. inspect the bracket

    d. turn them upside down periodically

96. You have responded to a motor vehicle accident on a busy city street. Your SOPs state you should disconnect the battery of the car in all wrecks. You open the hood and find high voltage warnings all over the engine compartment. Nothing in the engine compartment looks familiar as many engine items are missing. You should assume the car is _____.

    a. radioactive

    b. hydraulic

    c. electric

    d. hybrid

97. Which extremely toxic biological material is the easiest to make?

    a. Anthrax

    b. Ricin

    c. Tularemia

    d. RDD

98. You are checking your structural PPE and you discover a large rip in the vapor barrier of your coat. What should you do?

    a. Continue to use the PPE, but you should email your quartermaster that you will eventually need a replacement coat.

    b. Use vapor barrier adhesive to patch the liner and notify the department quartermaster.

    c. Replace the coat immediately.

    d. Ensure your duty uniform is worn beneath the jacket.

99. Firefighters are on the scene where two hazardous materials have escaped their containers. One of the materials is propane, with a vapor density of 1.6. The other material is methane, with a vapor density of 0.6. What can firefighters expect in this situation?

    a. The propane will rise and the methane will sink and collect at low points.

    b. Both the propane and methane will rise.

    c. The methane will rise and the propane will sink and collect at low points.

    d. Both the methane and propane will sink and collect at low points.

100. Many jurisdictions are working together to fight a wildfire on a mountain. The incident commander wishes to review the incident action plan. Who will she contact to get this information?

    a. The operations section chief

    b. The planning section chief

    c. The logistics section chief

    d. The finance/administration section chief

Phase III, Exam II: Answers to Questions

1. T	26. T	51. C	76. B
2. F	27. F	52. C	77. D
3. F	28. F	53. A	78. D
4. F	29. B	54. A	79. D
5. F	30. B	55. C	80. A
6. F	31. D	56. D	81. C
7. F	32. D	57. D	82. C
8. F	33. C	58. D	83. B
9. T	34. B	59. C	84. B
10. F	35. C	60. C	85. A
11. T	36. B	61. B	86. D
12. T	37. D	62. B	87. D
13. T	38. B	63. D	88. D
14. F	39. D	64. A	89. A
15. F	40. D	65. D	90. C
16. T	41. B	66. B	91. D
17. T	42. B	67. C	92. A
18. F	43. B	68. A	93. B
19. F	44. C	69. D	94. D
20. T	45. C	70. C	95. D
21. F	46. B	71. B	96. C
22. T	47. A	72. D	97. B
23. F	48. B	73. B	98. C
24. T	49. C	74. A	99. C
25. F	50. A	75. D	100. B

Phase III, Exam II:
Rationale & References for Questions

Question #1. Many fire apparatus are equipped with a technology that combines hearing protection/intercom/radio microphone into a single headset. NFPA 1001: 5.1.1.2. *FFHB, 2E:* Page 134.

Question #2. Specific chemical information is important for a safe and informed response. NFPA 472: 5.4.4; 5.5.1. *FFHB, 2E:* Page 817.

Question #3. Most have tempered glass, but a laminated safety glass and a polycarbonate glazing (Lexan) are common and can be found in these doors. NFPA 1001: 5.3.4; 6.3.2. *FFHB, 2E:* Page 523.

Question #4. First responders should understand the risk that EHS materials present. NFPA 472 4.2.1. *FFHB, 2E:* 746.

Question #5. Responders should understand the protective actions that are available to utilize. NFPA 472: 5.3.2 - 5.3.4. *FFHB, 2E:* Page 897.

Question #6. Figure 8-3 Older versions of fire extinguishers are labeled with colored geometrical shapes with letter designations. Ordinary combustibles are shown with a green triangle with the letter "A." NFPA 1001: 5.3.16. *FFHB, 2E:* Page 189.

Question #7. Responders should understand basic recognition of a dangerous BLEVE situation. NFPA 472: 5.2.2 - 5.2.4. *FFHB, 2E:* Page 783.

Question #8. Understanding protective clothing and its relationship to toxicology and health effects is important. NFPA 472: 5.3.3; 5.4.3. *FFHB, 2E:* Page 847.

Question #9. A group of firefighters is performing a fire safety inspection in a business. When they inspect the electrical hazards, they will check the fuse and breaker panels to verify that overcurrent protection devices have not been defeated and that all spaces on the panel are equipped with a breaker switch or blank to cover the opening. NFPA 1001: 5.5.1. *FFHB, 2E:* Page 672.

Question #10. First responders should have a basic awareness of the threat of terrorism and basic response actions. NFPA 472: 5.2.4. *FFHB, 2E:* Page 922.

Question #11. The type of fire, the characteristics of the burn, and the manner in which fuel is being supplied are the determining factors in deciding the best attack. NFPA 1001: 5.3; 6.3. *FFHB, 2E:* Page 94.

Question #12. Specific chemical information is important for a safe and informed response. NFPA 472: 4.2.1 - 4.2.2. *FFHB, 2E:* Page 818.

Question #13. A fire inspector from the Delmar Fire Department is conducting a fire prevention inspection in a new business. When he arrives, the business owner does not allow him to enter the building or conduct his inspection as planned. He will have to obtain an administrative warrant to enter the building without the consent of the business owner. NFPA 1001: 5.5.1. *FFHB, 2E:* Page 662.

Question #14. Understanding various protective actions is important for responder health and safety. NFPA 472: 5.2.2 - 5.2.4. *FFHB, 2E:* Page 866.

Question #15. In addition, a firefighter should never attempt to stand or don the SCBA unit while the apparatus is in motion. NFPA 1001: 5.3.2. *FFHB, 2E:* Page 162.

Question #16. Setting a knot is the finishing step of making sure the knot is snug in all directions of pull. NFPA 1001: 5.1.1.1. *FFHB, 2E:* Page 426.

Question #17. Cellar nozzles and Bresnan distributors can be used to fight localized fires in basements or cellars when firefighters cannot make a direct attack on the fire. NFPA 1001: 5.3 NFPA 6.3. *FFHB, 2E:* Page 286.

Question #18. Wet barrel hydrants have water in the barrel up to the valves of each outlet.They are used in areas that are not subject to freezing temperatures. NFPA 1001: 5.3.15. *FFHB, 2E:* Page 207.

Question #19. Fire hose needs to be tested prior to being placed in use and then retested annually during its lifetime. Hose also should be tested after being damaged and after repairs have been made. NFPA 1001: 6.5.3. *FFHB, 2E:* Page 274.

Question #20. Called a window search, this primary search takes advantage of speed by opening windows of rooms uninvolved and doing a quick look into the room. NFPA 1001: 5.3.9. *FFHB, 2E:* Page 465.

Question #21. When considering deployment methods for salvage covers, balloon toss deployments are done by two firefighters and shoulder toss deployments are done by one firefighter. NFPA 1001: 5.3.14. *FFHB, 2E:* Page 643-644.

Question #22. In the rehabilitation area, passive cooling practices (shade, air movement, rest) are less effective in reducing heat stress in firefighters than active cooling activities, such as forearm immersion technique. NFPA 1001: 5.3; 6.3. *FFHB, 2E:* Page 731.

Question #23. The other way to shut off the water flow from opened sprinklers is to shut down either the main sprinkler valve or a floor or sectional valve. NFPA 1001: 6.5.1. *FFHB, 2E:* Page 329.

Question #24. Depth of char is a term commonly used by fire investigators to describe the amount of time wooden material has burned. The deeper the char, the longer the material was burning or exposed to direct flame. NFPA 1001: 6.3.4. *FFHB, 2E:* Page 654-655.

Question #25. Here, depending on the time of day, the life hazard potential for fire victims is not as great, and the fire and life hazards to the firefighter can be greater. NFPA 1001: 5.3.10. *FFHB, 2E:* Page 615.

Question #26. Trusses create large void spaces. The obvious collapse danger with these void spaces is that the fire may be undetected while it continues to compromise the structural elements. NFPA 1001: 5.3.10; 6.3.2. *FFHB, 2E:* Page 362.

Question #27. Opening windows is the simplest way to open a compartment. Not every fire requires the removal of glass. NFPA 1001: 5.3.11. *FFHB, 2E:* Page 588.

Question #28. The sequence... or order of priority, is most generally as follows: 1.Those closest to the fire 2.Largest grouping of threatened people 3.Anyone else in the fire area 4.Those in areas that will eventually be exposed. NFPA 1001: 5.3.3. *FFHB, 2E:* Page 612.

Question #29. This may require using the floor below and advancing the hoseline toward the fire location. NFPA 1001: 6.5.1. *FFHB, 2E:* Page 334.

Question #30. A personal size-up usually includes an awareness of established work areas, hazardous energy, smoke conditions, and escape routes. Active cooling is something that is important once Firefighter Mason exits the structure into rehabilitation. NFPA 1001: 5.3; 6.3. *FFHB, 2E:* Page 730-731.

Question #31. A technique that has produced satisfactory results when operating off a flat roof beyond the reach of a ladder is to use a tool and a rope. NFPA 1001: 5.3.11. *FFHB, 2E:* Page 570.

Question #32. The running end is used for work such as hoisting a tool. NFPA 1001: 5.1.1. *FFHB, 2E:* Page 425.

Question #33. Smoke detectors are hard or permanently wired, battery operated, or a combination of hard wiring with a battery backup. NFPA 1001: 6.5.1. *FFHB, 2E:* Page 204.

Question #34. Testing should be conducted on fire hydrants periodically to ensure that they are operable and to determine the flow rate of the hydrants. NFPA 1001: 5.3.15. *FFHB, 2E:* Page 213.

Question #35. In general, anything that has a closed handle can be hoisted with a figure-eight or bowline, while longer cylindrical tools (i.e., ax and pike pole) can be hoisted using a clove hitch and half hitches. NFPA 1001: 5.1.1.1. *FFHB, 2E:* Page 449.

Question #36. The SCBA unit and protective equipment add weight and bulk to the firefighter, causing increased exertion with loss of body fluids through perspiration. These actions increase during firefighting operations. Firefighters must be aware of them and of the symptoms of heat stress and their own limitations and abilities. During rehabilitation, EMS personnel should monitor vital signs and firefighters must hydrate to replace body fluids. NFPA 1001: 5.3.1. *FFHB, 2E:* Page 168, 169.

Question #37. Pre-action systems are used in areas where the materials protected are of high value and water damage would be expensive, such as computer rooms and historical items. NFPA 1001: 6.5.1. *FFHB, 2E:* Page 321.

Question #38. Standpipe packs often use single-jacket hose to reduce the weight of the pack. NFPA 1001: 5.3. *FFHB, 2E:* Page 222.

Question #39. EP=NP+FL±E+SA NFPA 1001: 5.3 NFPA 6.3. *FFHB, 2E:* Page 92.

Question #40. Fire station tours should be tailored to fit the needs and interests of the visitors. The preschool children would be most receptive to a discussion on how to call for the fire department, as discussions about response procedures, detector placement, and first aid are beyond their comprehension level. NFPA 1001: 5.5.2. *FFHB, 2E:* Page 684.

Question #41. Prior to charging any lines, the static pressure is read on the main intake compound gauge.The pump operator then charges the first line with the desired volume, noting the pressure first.With this flow going, the operator again reads the intake gauge and gets the residual pressure or the remaining pressure left in the system after the flow and friction loss from the flow. NFPA 1001: 5.3.15. *FFHB, 2E:* Page 215.

Question #42. Career firefighters banded together near the beginning of the twentieth century to form a labor union, the International Association of Firefighters. This was not during the Civil War period. NFPA 1001: N/A. *FFHB, 2E:* Page 13-15.

Question #43. The line is brought to the entrance or landing as described earlier, charged, and advanced at the point where the attack is to be made. Care is taken to ensure that sufficient hose and personnel are available to move to the next floor without having to stop on the stairs. The nozzle is operated to darken down the fire at the next landing, then shut down, and quickly advanced to that level, where it is opened again to secure the landing and moved forward to maintain the floor. NFPA 1001: 5.3. *FFHB, 2E:* Page 255.

Question #44. However, picking the largest fire extinguisher available to put out a small fire can make it an expensive fire. NFPA 1001: 5.3.16. *FFHB, 2E:* Page 195.

Question #45. Ventilate as you advance, as long as it will not spread the fire. Ventilation allows the products of combustion to escape and provides a better environment. Check for outside openings such as windows and doors. This will provide a means of escape in an emergency and provide the firefighters' location to outside personnel. NFPA 1001: 5.3.1. *FFHB, 2E:* Page 150.

Question #46. This method, if used indoors, will usually greatly disturb the thermal balance, cause a loss of visibility in the structure as heat and products of combustion are circulated downward. When using this method, the firefighting team should attempt to shield themselves from the steam vapor by backing out of the area after applying the water. They can reenter as soon as the vapor begins to dissipate. NFPA 1001: 5.3.10. *FFHB, 2E:* Page 604.

Question #47. A very dense material will conduct heat better than a less dense material. NFPA 1001: 5.3; 6.3. *FFHB, 2E:* Page 88.

Question #48. In addition, 29 CFR 1910.134, NFPA 1404 and SCBA manufacturers' recommendations prohibit any facial hair, which may interfere with proper fit and seal of the face piece. NFPA 1001: 5.3.1. *FFHB, 2E:* Page 166.

Question #49. It is generally of little consequence in what order the occupancy is inspected as long as the method is efficient, systematic and thorough. Both firefighters can do the inspection in the order they want, as long as they cover everything and are systematic about their duties. This, of course, is subject to local standard operating procedures. A department can mandate that inspections be done in a certain pattern, in which case, that is the way all inspections must be performed. NFPA 1001: 5.5.1. *FFHB, 2E:* Page 662.

Question #50. It is a fact of nature that fire runs uphill faster than downhill.Caution:Weather is one of the main causes of firefighter fatalities on wildland fires. NFPA 1001: 5.3.19. *FFHB, 2E:* Pages 595 and 598.

Question #51. Hydraulic pistons are used on aerial ladder devices to raise and lower the ladder. They are not found on ground ladders. NFPA 1001: 5.3.6; 5.3.9; 5.3.10; 5.3.11; 5.5.3. *FFHB, 2E:* Pages 372-374.

Question #52. Attach locking-type pliers to the lock case and lock the jaws. The locking pliers must have a chain or rope attached so the firefighter can hold the pliers clear of the saw. A rotary saw with metal cutting blade is used to cut the lock shackles. Figure 17-50 Cut both sides of the shackle on one cut. NFPA 1001: 5.3.4; 6.3.2. *FFHB, 2E:* Pages 540-541.

Question #53. In addition to the aluminized fabric, proximity PPE features full face shields that are coated with an anodized gold material to help create a "mirrored" reflective surface. Without this special coating, the wearer could receive radiant facial burns, and the face shield could quite possibly melt. NFPA 1001: 5.1.1.2. *FFHB, 2E:* Page 130.

Question #54. As security requirements increase, home and business owners have begun to install many varieties of locks and security devices. If these types of devices are in use, the forcible entry team may need to find an alternate means of entry or use a rotary saw to gain access. NFPA 1001: 5.3.4; 6.3.2. *FFHB, 2E:* Page 530.

Question #55. The biggest problem with PASS devices results when wearers simply forget to turn their units on. This simple mental lapse has contributed to numerous firefighter fatalities. NFPA 1001: 5.1.1.2. *FFHB, 2E:* Page 135.

Question #56. A team of firefighters are placing an extension ladder to access a roof of a strip mall to check for fire extension and perform ventilation activities. The ladder should be placed above the roof level approximately five rungs. NFPA 1001: 5.3.6; 5.3.9; 5.3.10; 5.3.11. *FFHB, 2E:* Page 386.

Question #57. Pre-action systems are used in areas where the materials protected are of high value and water damage would be expensive, such as computer rooms and historical items. NFPA 1001: 6.5.1. *FFHB, 2E:* Page 321.

Question #58. Streetsmart Tip The rescuer must remember that none of these drags provides spinal immobilization and are intended to be utilized only in situations where greater harm will come to the patient if not immediately moved. NFPA 1001: 5.3.9. *FFHB, 2E:* Page 477.

Question #59. The firefighter must determine which way the door swings. The common way of describing this is in relationship to the forcible entry team. Doors with exposed or visible hinges will swing toward the forcible entry team. NFPA 1001: 5.3.4; 6.3.2. *FFHB, 2E:* Page 523.

Question #60. Observant firefighters on scene can provide fire investigators with reliable information to where the fire was, what spectators said about the fire, and how the fire reacted during fire attack. NFPA 1001: 6.3.4. *FFHB, 2E:* Page 651-654.

Question #61. Ropes constructed from polypropylene are primarily used for water rescue operations by the fire service. This is due to the fact that water has no effect on their strength, and even more beneficial is the fact that they will float. NFPA 1001: 5.1.1.1. *FFHB, 2E:* Page 421.

Question #62. Firefighter Taylor is correct. The range for effective span of control in the Incident Management System is 3 to 7 firefighters. If there are 22 total firefighters, they will need to be split into at least four divisions to properly accomplish an effective span of control. NFPA 1001: 6.1.1.1; 6.1.1.2. *FFHB, 2E:* Page 36.

Question #63. The top of a beam is subjected to a compressive force while the bottom of the beam is subjected to tension. NFPA 1001: 5.3.10; 6.3.2. *FFHB, 2E:* Page 346.

Question #64. There are certain behavioral characteristics and traits that are needed to operate successfully as a public safety telecommunicator. These include: the ability to perform multiple tasks; the ability to maintain composure in high-stress conditions; the ability to remember details and recall information easily; and the ability to exercise voice control. NFPA 1001: 5.2.1. *FFHB, 2E:* Page 48.

Question #65. Reports of emergencies can be received in a variety of ways, including: Conventional telephones Wireless or cellular telephones Emergency call boxes Automatic alarms TDD equipment Still alarms or walk-ups NFPA 1001: 5.2.1; 5.2.2. *FFHB, 2E:* Page 53-54.

Question #66. Additionally, a good fitting, tightly laced boot can help prevent ankle sprains and reduce foot fatigue. NFPA 1001: 5.1.1.2. *FFHB, 2E:* Page 132.

Question #67. Pressurized flammable liquids and gases are special fire hazards that should not be extinguished unless the fuel can be immediately shut off. The recommended method of fighting these fires is to turn off or disconnect the electrical power and then use an appropriate extinguisher, depending on the remaining fuel source. NFPA 1001: 5.3.16. *FFHB, 2E:* Page 186.

Question #68. Washing structural or wildland gear along with linens or other household items should be forbidden because cross-contamination can result. NFPA 1001: 5.1.1.2. *FFHB, 2E:* Page 136.

Question #69. Dump sites should be selected for availability to unload multiple tenders, turnaround area for the tenders, operational area, continued access to the fireground, and safety of personnel. NFPA 1001: 5.3.15. *FFHB, 2E:* Page 211.

Question #70. Spanner wrenches come in several sizes and are used to tighten or loosen couplings. NFPA 1001: 5.1.1.1. *FFHB, 2E:* Page 227.

Question #71. The point here is to avoid standing in front of any bumper after it has been heated. NFPA 1001: 5.3.7. *FFHB, 2E:* Page 600.

Question #72. Safety Firefighters should don all PPE necessary for the potential worst-case scenario. Granted, this approach may lead to "overdressing" for an incident. In these cases, the firefighters' company officer, incident commander, or incident safety officer may allow firefighters to "dress-down." NFPA 1001: 5.1.1.2. *FFHB, 2E:* Page 137.

Question #73. For a 1-A rating, the extinguisher should extinguish a wood crib fire of about one cubic foot.The ratings increase as the amount of fire suppressed increases. For instance, a 2-A extinguisher will put out twice the fire of a 1-A. NFPA 1001: 5.3.16. *FFHB, 2E:* Page 194.

Question #74. The Becket bend is utilized to tie ropes of equal diameter together, while the double Becket bend is used most often when tying ropes of unequal diameter. NFPA 1001: 5.1.1.1. *FFHB, 2E:* Page 429.

Question #75. Firefighter O'Malley is preparing for an upcoming presentation on fire safety to a high school class. Firefighter O'Malley should include current information in the presentation and should try to relate the subject matter to the audience. The presentation should include some "hands-on" activities, if possible, to reinforce learned concepts. NFPA 1001: 5.5.2. *FFHB, 2E:* Page 680.

Question #76. If it does freeze, it should be carefully moved to a place to thaw prior to any folding or bending. NFPA 1001: 5.3. *FFHB, 2E:* Page 223.

Question #77. Great success can be achieved with positive pressure ventilation or with the use of a hoseline stream to create negative pressure and pull the smoke out. NFPA 1001: 5.3.11. *FFHB, 2E:* Page 588.

Question #78. Purely defensive in design and execution, the trench cut is a roof opening that ventilates the cockloft area of a building where the fire is spreading under the roof. With the trench cut, the gases that are expanding horizontally under the roof are vented to the atmosphere and not permitted to pass a chosen defensive line. NFPA 1001: 5.3.11. *FFHB, 2E:* Page 579.

Question #79. A halyard, a pulley, and dogs (pawls, rung locks, ladder locks) are found on an extension ladder but not on a straight ladder. NFPA 1001: 5.3.6; 5.3.9; 5.3.10; 5.3.11; 5.5.3. *FFHB, 2E:* Pages 372-373.

Question #80. Head pressure measures the pressure at the bottom of a column of water in feet. Head pressure can be gained or lost when water is being pumped above or below the level of the pump. NFPA 1001: 5.3 NFPA 6.3. *FFHB, 2E:* Page 291.

Question #81. This method is not as quick as pulling the cylinder with a lock puller and should be used only if time allows. NFPA 1001: 5.3.4; 6.3.2. *FFHB, 2E:* Page 536.

Question #82. Based on the scene assessment, the first-arriving apparatus should be positioned to create a traffic barrier to help shield the greatest number of rescuers. NFPA 1001: 6.4.1. *FFHB, 2E:* Page 487.

Question #83. Hydrocarbons cover a wide range of substances in forms from gaseous to liquid to semisolid and solid. Common examples are heating oil, diesel fuel, gasoline, kerosene, petroleum jelly, paraffin, and asphalt. NFPA 1001: 5.3 NFPA 6.3. *FFHB, 2E:* Page 297.

Question #84. When the concentration of a gas falls into the range where it can ignite, it is said to be within its flammable limits. 3 is the only reading that would fall within the lower and upper flammable limits in this example. NFPA 1001: 5.3; 6.3. *FFHB, 2E:* Page 90-91.

Question #85. Open the valve of the cascade cylinder with the lowest pressure. This pressure must be higher than the pressure in the cylinder being filled. NFPA 1001: 5.3.1. *FFHB, 2E:* Page 178.

Question #86. Loosen the straps and rotate the SCBA unit under the arm along the rib cage as a first step. NFPA 1001: 5.3.1. *FFHB, 2E:* Page 170.

Question #87. The firefighter's next step in the assessment of the unconscious woman is to check for major bleeding. This is the last step in the initial assessment. Vital signs and focused histories and physical exams are performed after all the steps in the initial assessment are complete. NFPA 1001: 4.3. *FFHB, 2E:* Page 704.

Question #88. While doing company inspections, the engine company finds a building with special egress control devices. They can expect the building to have an approved fire detection and an approved fire sprinkler system. Also, the devices will deactivate with an activation of the fire sprinkler or fire detection system and with a loss of power to the building. NFPA 1001: 5.5.1. *FFHB, 2E:* Page 665.

Question #89. A work uniform that meets NFPA standards is not designed to protect the wearer from IDLH atmospheres, but can add another layer of reasonable protection under wildland, proximity, and structural ensembles. NFPA 1001: 5.1.1.2. *FFHB, 2E:* Page 136.

Question #90. Immediately exit the hazardous environment. All team members must exit. Never leave a firefighter alone. •If necessary, use the protective hood as a filter. Use the "buddy breathing" attachment on an SCBA as a last resort because this will deplete the air supply at least twice as fast. NFPA 1001: 5.3.1. *FFHB, 2E:* Page 171.

Question #91. Fire departments must establish an accountability system to track personnel entering an IDLH atmosphere. Firefighters must work in teams of two as a minimum. PASS devices must be activated. NFPA 1001: 5.3.1. *FFHB, 2E:* Page 168.

Question #92. Specific chemical information is important for a safe and informed response. NFPA 472: 4.2.1 - 4.2.2. *FFHB, 2E:* Page 809.

Question #93. First responders should have a basic awareness of the threat of terrorism and basic response actions. NFPA 472: 4.2.1. *FFHB, 2E:* Page 929.

Question #94. Signs and symptoms of critical incident stress include: Numbing and withdrawal Re-experiencing the event Flashbacks Depression Sleep difficulties Substance abuse Guilt Family problems Low job efficiency Fear of next incident Anger and hostility NFPA 1001: 6.1.1.1. *FFHB, 2E:* Page 116.

Question #95. Dry chemical or dry powder extinguishers carried on apparatus should occasionally be rotated upside down and shaken to keep powders from packing at the bottom as a result of vehicle vibration. NFPA 1001: 5.3.16. *FFHB, 2E:* Page 197.

Question #96. Instead, the firefighter will find high voltage warning labels on components. NFPA 1001: 5.3.7. *FFHB, 2E:* Page 600.

Question #97. First responders should have a basic awareness of the threat of terrorism and basic response actions. NFPA 472: 4.2.1. *FFHB, 2E:* Page 930.

Question #98. Both components rely on a layered protection system that includes a fire-resistive outer shell, vapor barrier, and thermal barrier. NFPA 1001: 5.1.1.2. *FFHB, 2E:* Page 128.

Question #99. Firefighters are on the scene where two hazardous materials have escaped their containers. One of the materials is propane, with a vapor density of 1.6. The other material is methane, with a vapor density of 0.6. The methane will rise and the propane will sink and collect at low points. NFPA 1001: 5.3; 6.3. *FFHB, 2E:* Page 87-88.

Question #100. The planning section chief is responsible for the development of the incident action plan. NFPA 1001: 6.1.1.1. *FFHB, 2E:* Page 39.

Phase Three, Exam Three

1. You are working with a rookie firefighter. You are teaching her to tie a bowline knot, and she just can't seem to catch on. You warn her that if she doesn't learn to correctly tie this knot, there is a good chance her knot will turn to a slip knot.

   a. True

   b. False

2. You are in staging and watching the scene of a structure fire. You see one of the attack lines suddenly jump off the ground. You can suspect water hammer.

   a. True

   b. False

3. A hard sleeve can be used to make an overflow dam.

   a. True

   b. False

4. The Hazard Communication standard 29 CFR 1910.1200 requires that MSDS be maintained for chemical safety.

   a. True

   b. False

5. You are preparing to enter the hallway of an apartment complex that has had a sprinkler activation. You have requested a "stop" as you know this is the only way to shut down water flow from opened sprinklers.

   a. True

   b. False

6. You have responded to a fire involving a lawn mower. When you spray water on it, the fire seems to explode. You decide to use Class D powder and quickly extinguish the fire. You know it is acceptable to sweep the powder up, place it in the original container, and store it back on the apparatus.

   a. True

   b. False

7. You are advancing toward a very large fire in a commercial business. You should ventilate by taking out the windows as you advance.

   a. True

   b. False

8.   You have responded to a commercial fire alarm in a local hardware/building supplies chain. The 2 A.M. alarm has used all available fire resources as there is another fire alarm in an apartment complex across town. The number one priority on both of these fire scenes is rescue/evacuation.

     a.   True

     b.   False

9.   You are part of a search and rescue team on a commercial structure with a glass entry door. You decide breaking the glass will be the fastest way to gain entry to the structure. However, you know the glass is laminated and you will need find another entry point.

     a.   True

     b.   False

10. You and your partner are conducting a primary search in a residential structure. Your assignment is to search an attached apartment. The vent crew has successfully vented above the seat of the fire and the attack crew has the fire knocked down. Visibility is extremely poor in the apartment you and your partner are searching. It is acceptable for you to take out windows as you advance your search.

     a.   True

     b.   False

11. While fighting a fourth floor residential apartment fire, you are assigned the task of advancing a line through a window. The safest manner to accomplish this is to advance the uncharged hoseline up the ladder.

     a.   True

     b.   False

12. You have responded to a multifamily, multistory residential structure fire in the downtown area of your district. The time is 2 A.M. and the weather is cold and snowy. The building is a care facility for the elderly. Upon arrival, you see smoke and flames coming from a fourth floor apartment. You see smoke coming from several of the other units on the same floor. The are occupants running from the building. A few of them have slipped and fallen on the icy sidewalk. Most are barefoot with no winter clothing and many are shivering. Most are exhausted from the rapid evacuation down the stairs. There are occupants on the balcony of the unit with flames showing and there are occupants on many of the balconies of the smoke-filled apartments. Your first priority would be to move the evacuated occupants to a warm environment.

     a.   True

     b.   False

13. You are a ventilation group supervisor and your job it to ventilate the roof of a four-story residential structure. Before you ascend the ladder, you should brief your crews of the location of the second roof exit.

     a.   True

     b.   False

14. While enjoying an off-duty afternoon in a movie theater, you find a seat burning in the second row. You tell your significant other to call 9-1-1 while you search for an extinguisher. You find four extinguishers with the older symbols on them. You know you want to pick the extinguisher with a yellow star and a letter inside it.

    a. True

    b. False

15. You are nozzle person on a dumpster fire. You are using a 1-3/4 attack line with a constant-pressure nozzle. You know that you alone control the gallonage.

    a. True

    b. False

16. For an efficient mobile water supply system to work well, you will need to have a water tender and a portable water tank.

    a. True

    b. False

17. As you advance up the stairwell of a multistory residential building with a fire alarm, you know that you will need hose to connect to the standpipe. You should not use the hose provided in the cabinet.

    a. True

    b. False

18. All deadly or harmful chemicals have a distinct odor and warning properties.

    a. True

    b. False

19. If you are in a rural area and decide to use a lake as a water source for a tender shuttle operation, you might make the operation easier if you keep your eyes open for a dry hydrant.

    a. True

    b. False

20. No matter whether you are using the over-the-head method or the coat method, the first three steps of checking the air supply, opening the cylinder, and comparing the cylinder gauge to the regulator gauge are always the same.

    a. True

    b. False

21. Combination headsets are considered adequate hearing protection.

    a. True

    b. False

22. A pressure tank being exposed to heat and flame is in danger of a BLEVE.

    a. True

    b. False

23. While in the hot desert city of Phoenix, Arizona, one might assume that the city fire hydrants are dry barrel hydrants.
    a. True
    b. False

24. You are fighting a room and contents fire in a residence. You are unsure if the room has been evacuated. You use a combination attack with coordinated ventilation to increase any victim chance of survival. This is a correct course of action.
    a. True
    b. False

25. Grasping the back plate and lifting the back plate/cylinder overhead to don an SCBA is called the coat method.
    a. True
    b. False

26. When you tactilely inspect a rope, you are visually inspecting the rope.
    a. True
    b. False

27. Out-of-service hose can be put back into service only if it is tested on its annual test due date.
    a. True
    b. False

28. An engine company arrives on the scene of a large storage building that is fully involved in flame with heavy smoke conditions. Bystanders in the next building report that the shed on fire is unoccupied and very little remained in storage in the facility. Using the concept of risk/benefit in this situation, you can assume that the team will make entry into the structure for search and rescue and an aggressive fire attack.
    a. True
    b. False

29. Firefighter Jong is searching for a new federal OSHA regulation about respiratory protection. Where would she look to find this federal regulation?
    a. United States Fire Administration manuals
    b. The Code of Federal Regulations
    c. National Fire Protection Association standards
    d. The National Institute for Occupational Safety and Health reports

30. You are called to a boat fire on a lake. Upon arrival, you find a large wooden boat burning.The boat is docked. There are no immediate fire hydrants visible. You should _____.
    a. use a reverse lay to the nearest hydrant
    b. request a water tender
    c. use booster tank water sparingly
    d. consider drafting

31. You are on a rescue scene. You are attempting to extend a rope for a rescue operation, as it is too short. You know that you have a duplicate rope in your apparatus. You know that you should consider the _____ to assist your operation.
    a. Becket bend
    b. bowline
    c. water knot
    d. figure-eight

32. When you are using a BC fire extinguisher, you know _____.
    a. the agent and nozzle will not conduct electricity
    b. the agent will not conduct electricity, but the nozzle will
    c. the nozzle will not conduct electricity, but the agent will
    d. you are using the wrong extinguisher

33. You are checking your structural PPE and you discover a large rip in the vapor barrier of your coat. What should you do?
    a. Continue to use the PPE, but you should email your quartermaster that you will eventually need a replacement coat.
    b. Use vapor barrier adhesive to patch the liner and notify the department quartermaster.
    c. Replace the coat immediately.
    d. Ensure your duty uniform is worn beneath the jacket.

34. Which of the following statements about using communications equipment is false?
    a. Clipping of a message can only occur in the beginning of a message and not at the end.
    b. Microphones and portable radios should be placed in their appropriate storage location or in a protected location.
    c. Proper radio discipline is important during active incidents.
    d. Radio transmissions should be within the time parameters established by the department or local jurisdiction.

35. You are preparing to hoist a chainsaw to the roof of a three-story building. Which of these knots is acceptable for this operation?
    a. clove hitch
    b. Becket bend
    c. figure-eight follow-through
    d. rescue knot

36. You respond to a reported smoke detector activation in the mainframe computer room of a school. Upon arrival, you are told there is a sprinkler system, but it has not activated. Based on this information, you can be relatively sure that the system is a _____ system.
    a. deluge
    b. dry pipe
    c. wet pipe
    d. pre-action

37. You are assigned to the search team of a two-story residential structure fire with heavy smoke showing. The neighbors report that the homeowner is probably inside. There is very heavy yellowish color smoke coming from the eves of the roof. It appears to blow out and then is sucked back in. You are attempting to force the door to conduct a search for the homeowner. You know you must ensure _____.

    a. coordination with other teams

    b. the neighbors on the opposite side have not seen the homeowner leave

    c. you take oxygen with you as the patient will likely need it

    d. the door is locked before you force it open

38. Firefighter O'Malley is preparing for an upcoming presentation on fire safety to a high school class. Which of the following is not important for Firefighter O'Malley to consider in preparing for the presentation?

    a. The presentation must include current information.

    b. The presentation should try to relate the subject matter to the audience.

    c. The presentation should include some "hands-on" activities, if possible, to reinforce learned concepts.

    d. The presentation should end with a comprehensive test to check that the audience was paying attention.

39. A group of preschool children have come to visit the fire station. Firefighter Williams is giving a tour of the facility. Which of the following things should be focused on during his presentation to the preschool children?

    a. How to call for the fire department in an emergency

    b. Fire department response procedures

    c. Proper placement of carbon monoxide detectors

    d. Upcoming first aid classes

40. You are explaining hydraulics to your rookie firefighters. You explain the formula for calculating engine pressure. One of the students does not understand the formula as you explain it. Therefore, you write it on the board as _____.

    a. $EP=NP-FL\pm E+SA$

    b. $EP=NP*FL\pm E+SA$

    c. $EP=NP/FL\pm E+SA$

    d. $EP=NP+FL\pm E+SA$

41. As a conscientious firefighter, you realize that dynamic influences can reduce the TPP of your structural gear. Fabric compression is one of these factors resulting from _____.

    a. folding your structural pants and placing them under your helmet while the gear is in storage

    b. wearing webbing that carries an axe and a flashlight

    c. wearing an SCBA

    d. all of the above

42. What is a common choking agent that is found in most communities?

    a. Nerve agents

    b. Chlorine

    c. Sarin

    d. Anthrax

43. How can a firefighter assist fire investigators in discovering how and why a fire occurred?

    a. Remove all of the contents of the burned rooms to the outside of the structure for ease in conducting investigations.

    b. Use large volumes of water to ensure materials are thoroughly soaked prior to fire investigation efforts.

    c. Being observant of the variety of conditions and occurrences on the fire scene and passing this information to fire investigators.

    d. All of the above are correct.

44. You respond to a reported smoke detector activation in the diorama room of a museum. Upon arrival, you are told there is a sprinkler system, but it has not activated. Based on this information, you can be relatively sure that the system is a _____ system.

    a. deluge

    b. dry pipe

    c. wet pipe

    d. pre-action

45. You are flighting a wood pile fire at a cabinet factory. The fire is relatively small. You estimate it to be 1 cubic foot. The wind is not blowing, the temperature is 50°F, and the fire is not increasing in size rapidly. The nearest extinguisher is rated as a 30A160BC. There is a 2A 10BC extinguisher mounted on the wall 100 feet away. What should you do?

    a. Use the 30A160BC extinguisher to put out the fire.

    b. Wait for an attack line from the engine.

    c. Watch the fire while you send your partner for the 2A10BC extinguisher.

    d. Look for a Class A only extinguisher.

46. You are working a multifamily, multistory structure fire that is in defensive mode. You are assigned the task of putting water in an eighth story window. You know that you will need to deploy _____.

    a. a 2-1/2 inch line and a straight ladder

    b. a master stream

    c. a 1-3/4 inch line and a solid tip nozzle

    d. an LDH attack line

47. You are in an IDLH atmosphere and your hear the order to evacuate on another firefighter's radio. You discover your radio is not operating. You have become disoriented and your partner has left you. You yell to gain anybody's attention and nobody responds._____ should alert IC to your absence.

    a. Your partner

    b. Your PASS device

    c. The accountability system

    d. All of the above

48. You are searching for the seat of a fire with your partner in the interior of a residential structure fire. As you advance down a hallway, you find you are following an exterior wall with windows and a door. The products of combustion are thick and visibility is just inches. However, you do not feel an increase in ambient temperature. Your course of action should be to _____.

    a. retreat and find other access to the fire

    b. notify IC and retreat to find other access to the fire

    c. break out windows as you advance

    d. hold your position at the mouth of the hallway and have your partner advance toward the fire

49. While doing an interior search of a commercial structure fire, your partner's low-air alarm activates. He notifies you that he needs to leave. You both retreat and follow your search pattern out of the building. About 200 feet from the exit, your partner notifies you he is out of air. You see him remove his regulator, pull his hood over the opening, and put his face to the floor. You should _____.

    a. stop, notify IC that you are in trouble, and buddy breath until the "two out" team can rescue you

    b. start buddy breathing and continue your emergency egress

    c. maintain contact with your partner and continue your emergency egress

    d. continue your emergency egress with the intent of sending in more air for him if you lost him on the way out of the building

50. You are responding to a reported structure fire in a rural area outside of town. Your officer calls for water tenders to respond. She tells you the hydrants in this area have almost no pressure. She knows this from hydrant _____.

    a. size

    b. testing

    c. type

    d. shape

51. You are in staging on a hazardous materials call. The hazmat technicians are suiting up for entry into a commercial structure in which you used to work. In your estimation, you calculate that the room they need to access is about 5 minutes into the building. You see them donning open-circuit supplied air respirators. You know the cylinders they will be carrying only contain 5 to 10 minutes of air. You should _____.

    a. immediately notify IC

    b. immediately notify staging

    c. start the timer on your watch and monitor the situation

    d. do nothing

52. You are trying to put a swift water rescue throw bag together to keep in your bunker coat pocket. Your best choice, in this case, would be to use _____ rope.

    a. polyester

    b. polypropylene

    c. polyethylene

    d. nylon

53. On a residential structure fire scene in the winter, you notice that one of the attack lines has frozen. You should _____.

    a. heat it to thaw the water

    b. carefully move it to a place where it can thaw

    c. roll it and place it on the apparatus

    d. discard it

54. You are working in a station that responds to wildland fires. You are folding the station laundry from the clothes dryer and you find your partner's wildland bandana in with the station bathroom towels. What should you do?

    a. Re-wash the towels to ensure no cross-contamination with products of combustion, and remind your partner of the hazards of this practice.

    b. Tell your partner that practice is not acceptable and in the future to keep his/her PPE separate from station laundry. Fold and put away the laundry.

    c. Dispose of the station bathroom towels.

    d. Put away the laundry and say nothing.

55. While fighting a wildland fire, you notice you and your crew have hiked onto a saddle above the advancing fire. It is 3 P.M. The weather is changing rapidly. Currently the wind is blowing the fire back down slope. You should _____.

    a. evacuate to a safe zone

    b. notify the plane of the weather change

    c. deploy your shelter

    d. establish a watch to monitor the weather

56. You have been charged with search and rescue of a doctor's office with a fire alarm activation but no smoke or flames showing. You have responded to this location numerous times for fire alarms, and they have all proved to be false. You and your partner prepare to enter the structure. You arrive at the entry door, and see through the window there is no fire or smoke, and no evidence of any damage. You do not see any occupants. Your first action is to check to see if _____.

    a. there are auxiliary locking devices

    b. there is a handle or knob on the outside of the door

    c. there are exposed hinges

    d. your partner has the K tool

57. You have responded to a motor vehicle accident on a busy city street. Your SOPs state you should disconnect the battery of the car in all wrecks. You open the hood and find high voltage warnings all over the engine compartment. Nothing in the engine compartment looks familiar as many engine items are missing. You should assume the car is _____.

    a. radioactive

    b. hydraulic

    c. electric

    d. hybrid

58. You are fighting a very large and hot room and contents fire in a residential structure.Because of poor visibility, your team decides to make a(n) _____.After this tactic, your team will probably have to back out of the fire area until the steam is dissipated.

    a. transitional attack

    b. indirect attack

    c. direct attack

    d. fire attack

59. _____ is an important concept for firefighters because it affects how to enter and function in a room or area that is on fire or where the fire has just been extinguished: firefighters must stay low to the floor in attacking structural fires.

    a. Backdraft

    b. Rollover

    c. Thermal layering

    d. Thermal conductivity

60. Your engine is attached to the hydrant using two 2-1/2 inch supply lines. You notice that one of the lines is leaking a considerable amount of water at one of the couplings. You should _____.

    a. pull another section of 2-1/2 inch line for replacement

    b. tighten the coupling with a hydrant wrench

    c. tighten the coupling with spanner wrenches

    d. use a hose bridge

61. Once you don your face piece, you should check for a proper seal by
_____.

    a. placing your hand over the regulator hole and attempting to inhale

    b. attaching the regulator, taking a breath and holding it, and listening and feeling for any air leaks

    c. smelling for smoke or fumes once you enter the IDLH

    d. No check is needed. It is mandated by NFPA that all masks be custom fit.

62. What has changed to bring air monitoring devices to the first responder level?

    a. Technology

    b. Size of fire apparatus

    c. Types of chemicals

    d. Numbers of firefighters

63. Firefighters from an interior fire attack team are arriving in the rehabilitation area. When providing nourishment to the firefighters, what is the ideal balance of protein, fat, and carbohydrates to stabilize the release of insulin into the bloodstream?

    a. 30% protein, 30% fat, 40% carbohydrates

    b. 30% protein, 50% fat, 20% carbohydrates

    c. 50% protein, 50% fat, no carbohydrates

    d. 10% protein, 60% fat, 30% carbohydrates

64. You are a firefighter on scene of a car fire. There are flames showing from the grill and heavy smoke is billowing out from the wheel wells and from under the car. You watch two firefighters on a hoseline advance toward the engine compartment fire. They are advancing in a straight line directly in front of the car toward the grill. You should have them _____.

    a. wait until you open the hood

    b. approach from the side

    c. approach from the back

    d. wait until foam can be applied

65. Which of the following is not true regarding hydration of firefighters in rehabilitation areas?

    a. Sports drinks are best diluted 50 percent with water to speed their absorption into the system.

    b. Juice or sports drinks are the best choice for hydration in the first hour of activities.

    c. Sports drinks can help in replacing electrolytes and nutrients along with water.

    d. A good rule of thumb for firefighters to strive for during periods of work is to drink a quart of water an hour.

66. During the daily check of your SCBA on your engine, you notice that the cylinder and regulator gauges differ. The cylinder gauge reads full, but the regulator gauge reads 1/2. You should _____.

    a. immediately fill the bottle until the regulator gauge reads full

    b. tag the SCBA, but as it is your unit, you should go ahead and use it for the shift, as you know the bottle is full

    c. replace the bottle and check the status of the gauges

    d. remove the pack from service

67. You are a new firefighter assigned to a station that handles technical rescue for the fire district. You would like to customize your structural trousers with sewn-in webbing. You can have this customizing done _____.

    a. at any tent repair store or other store that deals with heavy canvas

    b. by an NFPA-certified manufacturer

    c. by your wife, if she has a heavy-duty sewing machine

    d. by nobody

68. Your department practice is to issue PASS devices to individual firefighters. You are donning PPE and realize you left your PASS device at the station. What should you do?

    a. Continue to don PPE and prepare for initial fire attack and say nothing about your missing PASS device.

    b. Realize that you will have to resort to the back-up PASS practice of screaming loudly if you get into trouble.

    c. Realize that not wearing a PASS device is as serious as not turning one on, and you should notify command immediately.

    d. Take a PASS device from another firefighter on scene and owe him or her a big favor.

69. You are venting the flat roof of a six-story apartment building and your chainsaw fails. You are unable to restart it. The attack crews are staged and waiting for ventilation. Under direction of IC and, if possible, you should _____.

    a. advise attack crews to begin the assault on the fire; ventilation is not possible

    b. use a rope and a tool to ventilate the windows on the sixth floor

    c. use a knife or other tool to begin removing the roofing cover

    d. use a rope and a tool to ventilate the windows on the fire floor

70. An important agent maintenance duty for dry chemical extinguishers is to _____.

    a. wash the exterior with soap and water

    b. visually examine the exterior for soap residue

    c. inspect the bracket

    d. turn them upside down periodically

71. An incident commander is splitting up the firefighters on the scene of a structure fire in a one-story house. He decides to send four firefighters and one officer to be responsible for search and rescue. What incident command designation will he give the four firefighters and one officer?

    a. Search and rescue division

    b. Search and rescue group

    c. Search and rescue crew

    d. Search and rescue strike team

72. Several firefighters are discussing the growth stage of fire. The speed of the growth and ultimately the size of the fire are dependent on which of the following factors?

    a. Oxygen supply

    b. Fuel

    c. Container size

    d. All of the above

73. Many units from the Delmar Fire Department are on the scene of a fire in an apartment building. The incident commander wants to check on the availability of getting additional hose from the fire department storage area to use on the scene. Who should he contact?

    a. The operations section chief

    b. The planning section chief

    c. The logistics section chief

    d. The finance/administration section chief

74. A group of firefighters are discussing the storage of ladders. Which of the following is incorrect about ladder storage?

    a. Ladders should be stored in clean, dry places.

    b. Ladders should be stored with one support point only.

    c. Ladders should be placed in a storage area where they will not be subjected to heat or exhaust.

    d. Ladders should not be placed in out-of-sight areas where they can be a tripping hazard.

75. You are in the stairwell of an apartment complex. There is smoke showing from the fourth floor. You are unsure where the fire is on the fourth floor, but know the first through third floors are clear. You decide to make the standpipe connection of your high rise pack on the _____ floor

    a. second

    b. third

    c. fourth

    d. fifth

76. You are tasked with fire attack of a residential room and contents fire. It is a snowy wintry night and the temperature is hovering around 28 degrees. The ventilation crews confirm that a roof hole close to the seat of the fire is complete. You see that there is not much relief from the heat and smoke. You should consider _____.
    a. a retreat
    b. a direct attack
    c. an indirect attack
    d. hydraulic ventilation

77. When using a portable fire extinguisher, you know it is a good idea to _____.
    a. use the largest extinguisher available
    b. start fighting the smoke first
    c. quickly locate a second extinguisher
    d. look for an exit way behind you

78. Firefighter Lopez is inspecting an extension ladder. Which of the following is not true about her inspection of the ladder?
    a. The halyard should be inspected for undue wear.
    b. The heat sensor label should be checked for presence and condition.
    c. The ladder should be examined for any discoloration from heat damage, cracks, and deformed areas.
    d. All of the above are true.

79. In the 1970s, the US government commissioned a panel of fire service experts to study the country's changing fire problem. As a result of the document this group published, many positive changes occurred in fire service at various levels of government. What was the name of the document that was produced?
    a. America Burning
    b. The National Commission on Fire Protection and Control Journal
    c. The United States Fire Administration Report
    d. Fire Protection and Control in the US

80. As an active firefighter, you should realize that PPE provides you the _____ level of protection and should be considered the _____ of protection.
    a. minimum / last resort
    b. maximum / front line
    c. minimum / front line
    d. maximum / last resort

81. You are in mop-up mode on a structure fire where you used a considerable amount of foam. The driver/operator of the engine tells you the foam system on the apparatus uses an around-the-pump proportioner. You should expect to _____.
    a. deplete the entire foam capacity
    b. back-flush the pump
    c. prime the pump
    d. discard the first section of hose

82. The ACGIH recommends that you only be exposed to 20 ppm of a chemical; while OSHA requires that the exposure be less than 35 ppm. To which one do you have to be legally adhered?

    a. OSHA

    b. ACGIH

    c. NIOSH

    d. None of the above

83. Two firefighters are performing overhaul operations inside a structure. Which of the following sets of tools would be most useful to the firefighters?

    a. Salvage covers, a hammer, and nails

    b. Bolt cutters and sprinkler wedges

    c. Squeegees and a water vacuum

    d. A carry-all, pike poles, and axes

84. While doing an interior attack in a residential structure, you and your partner find it necessary to make your own escape from an interior bedroom. You decide your best course of action is to breech the wall and escape through an adjacent bedroom. You and your partner punch a hole in the drywall and find a safe egress. Your partner, who is smaller than you, has a bit of trouble fitting through the breech, and you decide you will have more difficulty. Your best course of action is to _____.

    a. wait for your partner to escape and go for help

    b. remove your SCBA and proceed through the breech while holding the SCBA in front of you

    c. attempt to breech another hole in a different location

    d. loosen the straps and rotate the SCBA under your arm and attempt to fit through the breech

85. You have just received your new wildland PPE ensemble. As you are unpacking it, you realize you have questions about the care and limitations of use of the helmet. You should_____.

    a. check NFPA 1001

    b. check the information that came with the PPE

    c. check the tag inside the helmet

    d. ask your officer

86. The NFPA 1500 standard covers a multitude of topics related to safety of fire service personnel. Which of the following is not a topic covered by NFPA 1500?

    a. Training and education

    b. Personnel accountability procedures

    c. Protective clothing and equipment

    d. Critical incident stress program

87. "Black fire" is apparent when a team of firefighters prepares to initiate fire attack. The smoke is high volume with a turbulent velocity and is ultradense and black. What should the firefighters expect to occur when they "read" this smoke condition?

   a. Backdraft

   b. Flash point

   c. Autoignition and flashover

   d. The decay phase

88. You have responded to a residential structure fire in a neighborhood that is known for its high crime rates. It is 2 A.M. and there is a car parked in front of the home on the street. The home has bars over the windows. You are on the search team and you attempt to force the front door for entry. You spend a minute or so attempting to force the door and it simply will not move. You _____.

   a. suspect an auxiliary locking device

   b. should keep trying to force the door

   c. should try pulling the lock

   d. should try using a K tool

89. A firefighter is treating a patient who has been involved in an automobile accident. The firefighter checks the patient for signs and symptoms of internal bleeding. Which of the following is not an indication of possible internal bleeding?

   a. Cold and clammy skin

   b. Red, flushed skin

   c. Dilated pupils

   d. Bruising of the skin

90. You are securing your life-safety rope around an anchor and are preparing to rappel down a 10-foot drop off. There are no trees or other suitable anchors in the area. You decide to use the axle of your apparatus. You use which knot to accomplish this anchor?

   a. figure-eight follow-through

   b. Becket bend

   c. clove hitch

   d. half hitch

91. You respond to a possible residential structure fire call and find that a teenager has burned popcorn in the microwave oven. There is no fire, but the smell of burnt popcorn smoke is throughout the first floor of the house. As a customer service, you should consider _____.

   a. taking the microwave to the dumpster

   b. PPV of the unaffected portions

   c. PPV of the affected portions

   d. removing all popcorn so the episode cannot be repeated

92. You have been working the interior of a residential structure fire and find yourself low on air. You and your partner retreat through the garage and you find the door will not raise. The power has been disconnected to the house so the automatic garage door opener will not work. The easiest method of exit is to _____.

    a. take out a window on the door to create an exit

    b. take out a panel of the door with your axe

    c. make a door in the garage wall

    d. pull the release rope

93. You grab a section of 1-3/4 inch hose that is rolled and stored in your engine. You are going to connect it to one of the discharges of the engine. Usually this type of hose is rolled with the _____ end inside.

    a. male

    b. female

    c. other

    d. outside

94. Firefighters arrive on a scene to find a Class C fire. Which of the following methods should the firefighters use to best control this type of fire?

    a. Cooling the fire with large quantities of water

    b. The application of a smothering agent to prevent oxygen from getting to the fuel

    c. Removing the flow of electricity and then re-analyzing the class of the fire for further actions

    d. Applying a dry powder to stop the chemical reaction

95. You respond to a fuel station with a propane pump burning. Your first action should be to _____.

    a. pull a hoseline

    b. grab a Class B extinguisher

    c. move the closest vehicle away

    d. shut off or have the pumps shut off

96. Firefighter Hernandez is getting ready to transmit a message on a mobile radio to the emergency communications center. Which of the following is false concerning the transmission of his message?

    a. The transmission must be within the time parameters established by the department or local jurisdiction.

    b. When he is ready, Firefighter Hernandez can hit the push-to-talk button and begin speaking immediately.

    c. Firefighter Hernandez should make sure the information he is transmitting is accurate, clear, and complete.

    d. Firefighter Hernandez should not attempt to transmit messages by radio while eating or chewing anything.

97. You and your partner are searching a commercial structure next door to a working structure fire. There is no smoke or fire in this building. You find a victim who appears to have fallen from a second story balcony. The appropriate rescue carry for this patient is _____.

    a. the firefighter's drag

    b. the webbing sling drag

    c. the clothing drag

    d. none of the above

98. A firefighter is checking a wall inside a structure for signs of hidden fire. Which of the following would indicate a hidden fire in the wall?

    a. The wall feels hot to the touch.

    b. A burning smell is noted near the outlets on the wall.

    c. The sound of items burning can be heard when near the wall.

    d. All of the above are signs of hidden fire in the wall.

99. Firefighters arrive on a scene to find a Class A fire. Which of the following methods should the firefighters use to best control this type of fire?

    a. Cooling the fire with large quantities of water

    b. Applyinga smothering agent to prevent oxygen from getting to the fuel

    c. Removing the flow of electricity and then re-analyzing the class of the fire for further actions

    d. Applying a dry powder to stop the chemical reaction.\

100. Which of the following statements about steel is incorrect?

    a. Steel is a mixture of carbon and iron ore heated and rolled into structural shapes.

    b. Steel has poor tensile, shear, and compressive strength.

    c. Cooling structural steel with fire streams is just as important as attacking a fire in the structure.

    d. Cold drawn steel loses its strength at lower temperatures than extruded structural steel.

Phase III, Exam III: Answers to Questions

1. T	26. F	51. D	76. D
2. T	27. F	52. B	77. D
3. T	28. F	53. B	78. D
4. T	29. B	54. A	79. A
5. F	30. D	55. A	80. A
6. F	31. A	56. C	81. B
7. T	32. A	57. C	82. A
8. F	33. C	58. B	83. D
9. F	34. A	59. C	84. D
10. T	35. C	60. C	85. B
11. T	36. D	61. B	86. B
12. F	37. A	62. A	87. C
13. T	38. D	63. A	88. A
14. F	39. D	64. B	89. B
15. F	40. D	65. B	90. A
16. T	41. D	66. C	91. C
17. T	42. B	67. A	92. D
18. F	43. C	68. C	93. A
19. T	44. D	69. D	94. C
20. T	45. C	70. D	95. D
21. T	46. B	71. B	96. B
22. T	47. D	72. D	97. D
23. F	48. C	73. C	98. D
24. T	49. C	74. B	99. A
25. F	50. B	75. B	100. B

Phase III, Exam III:
Rationale & References for Questions

Question #1. If left on the outside, there is a much greater chance of it getting caught and the knot inverting to a slip knot. NFPA 1001: 5.1.1.1. *FFHB, 2E:* FFHB: Page 432.

Question #2. Damage from improper actions by firefighters include opening and closing valves or hydrants too quickly (thus creating a water hammer), failing to open or close a hydrant or other valve fully, or cross-threading the threads on the connections. NFPA 1001: 5.3.15. *FFHB, 2E:* FFHB: Page 216.

Question #3. Responders should understand the protective actions that are available to utilize. NFPA 472: 5.3.2 - 5.3.4. *FFHB, 2E:* FFHB: Page 895.

Question #4. Specific chemical information is important for a safe and informed response. NFPA 472: 4.2.1 - 4.2.2. *FFHB, 2E:* FFHB: Page 818.

Question #5. The other way to shut off the water flow from opened sprinklers is to shut down either the main sprinkler valve or a floor or sectional valve. NFPA 1001: 6.5.1. *FFHB, 2E:* FFHB: Page 329.

Question #6. Personnel must use the correct and uncontaminated (clean, dry, and without any other foreign materials in it; basically as it comes from the factory) extinguishing agent for each different Class D material. NFPA 1001: 5.3.16. *FFHB, 2E:* FFHB: Page 186.

Question #7. When venting a series of windows, the firefighter must work toward the escape point. NFPA 1001: 5.3.11. *FFHB, 2E:* FFHB: Page 581.

Question #8. Here, depending on the time of day, the life hazard potential for fire victims is not as great, and the fire and life hazards to the firefighter can be greater. NFPA 1001: 5.3.10. *FFHB, 2E:* FFHB: Page 615.

Question #9. Most have tempered glass, but a laminated safety glass and a polycarbonate glazing (Lexan) are common and can be found in these doors. NFPA 1001: 5.3.4; 6.3.2. *FFHB, 2E:* FFHB: Page 523.

Question #10. While inside a structure, firefighters may "vent as they go" to help relieve smoke pressure, but only if such will not cause significant fire spread. NFPA 1001: 5.3.9. *FFHB, 2E:* FFHB: Page 465.

Question #11. The best and safest manner is to advance an uncharged hoseline up the ladder and into the building or onto a fire escape before it is charged. NFPA 1001: 5.3. *FFHB, 2E:* FFHB: Page 257.

Question #12. The sequence... or order of priority, is most generally as follows: 1.Those closest to the fire 2.Largest grouping of threatened people 3.Anyone else in the fire area 4.Those in areas that will eventually be exposed. NFPA 1001: 5.3.3. *FFHB, 2E:* FFHB: Page 612.

Question #13. There should always be two easily recognizable ways off a roof. NFPA 1001: 5.3.11. *FFHB, 2E:* FFHB: Page 581.

Question #14. Figure 8-3 Older versions of fire extinguishers are labeled with colored geometrical shapes with letter designations. Ordinary combustibles are shown with a green triangle with the letter "A." NFPA 1001: 5.3.16. *FFHB, 2E:* FFHB: Page 189.

Question #15. The automatic or constant-pressure nozzle has a flow that can be adjusted by the pump operator, who increases the pressure, which in turn increases the gallons flowing. NFPA 1001: 5.3 NFPA 6.3. *FFHB, 2E:* FFHB: Page 283.

Question #16. Tenders combined with portable water tanks can efficiently provide large volumes of water to a fireground operation. NFPA 1001: 5.3.15. *FFHB, 2E:* FFHB: Page 205.

Question #17. Class I systems are designed for use by the fire department or trained personnel such as a fire brigade.No hose is provided for this class. NFPA 1001: 6.5.1. *FFHB, 2E:* FFHB: Page 329.

Question #18. Understanding various protective actions is important for responder health and safety. NFPA 472: 5.2.2 - 5.2.4. *FFHB, 2E:* FFHB: Page 28.

Question #19. A dry hydrant is not really a fire hydrant, but a connection point for drafting from a static water source such as a pond or stream, not a pressurized one. They are used primarily in rural areas with no water system. They may be found in urban and suburban areas as a backup or where certain buildings may be of great distance from other water sources, but a pond or stream is located nearby. NFPA 1001: 5.3.15. *FFHB, 2E:* FFHB: Page 208.

Question #20. Donning Self-Contained Breathing Apparatus, Over-the-Head Method 1.Check the air supply. 2.Slowly open the cylinder valve. 3.Compare the cylinder gauge to the regulator gauge. Donning Self-Contained Breathing Apparatus, Coat Method 1.Check the air supply. 2.Slowly open the cylinder valve. 3.Compare the cylinder gauge to the regulator gauge. NFPA 1001: 5.3.1. *FFHB, 2E:* FFHB: Page 159, 160.

Question #21. Many fire apparatus are equipped with a technology that combines hearing protection/intercom/radio microphone into a single headset. NFPA 1001: 5.1.1.2. *FFHB, 2E:* FFHB: Page 134.

Question #22. Understanding of basic chemical and physical properties is important for the health and safety of emergency responders. NFPA 472: 5.2.2 - 5.2.4. *FFHB, 2E:* FFHB: Page 782.

Question #23. Wet barrel hydrants have water in the barrel up to the valves of each outlet.They are used in areas that are not subject to freezing temperatures. NFPA 1001: 5.3.15. *FFHB, 2E:* FFHB: Page 207.

Question #24. Ventilation with this combination attack controls the flow of fire gases and steam, improving the survival chances of victims, and makes this attack the type typically used by firefighters in structural firefighting. NFPA 1001: 5.3 NFPA 6.3. *FFHB, 2E:* FFHB: Page 290.

Question #25. Job Performance Requirement 7-1 Over-the-Head Method Grasp the back plate with both hands. Lift the back plate/cylinder overhead. NFPA 1001: 5.3.1. *FFHB, 2E:* FFHB: Page 154.

Question #26. They should also be inspected tactilely by running the rope through your hands. NFPA 1001: 5.1.1.1. *FFHB, 2E:* FFHB: Page 440.

Question #27. Fire hose needs to be tested prior to being placed in use and then retested annually during its lifetime. Hose also should be tested after being damaged and after repairs have been made. NFPA 1001: 6.5.3. *FFHB, 2E:* FFHB: Page 274.

Question #28. In this situation, when applying the concept of risk/benefit, you can assume that the firefighters will not take significant risks to their safety to save what is already lost. This is especially true in light of the reports that there are no victims inside and very little property to save. NFPA 1001: 5.3; 6.3. *FFHB, 2E:* FFHB: Page 730.

Question #29. Firefighter Jong would need to look in the Code of Federal Regulations, which is a set of documents that includes all federally promulgated regulations for all federal agencies. NFPA 1001: 6.1.1.1. *FFHB, 2E:* FFHB: Page 109.

Question #30. Other natural sources of surface water are rivers, lakes, and ponds. Drafting The pumping of water from a static source by taking advantage of atmospheric pressure to force water from the source into the pump NFPA 1001: 5.3.15. *FFHB, 2E:* FFHB: Page 204.

Question #31. The Becket bend is utilized to tie ropes of equal diameter together, while the double Becket bend is used most often when tying ropes of unequal diameter. NFPA 1001: 5.1.1.1. *FFHB, 2E:* FFHB: Page 429.

Question #32. Class C extinguishers have extinguishing agents and hoses with nozzles that will not conduct electricity. NFPA 1001: 5.3.16. *FFHB, 2E:* FFHB: Page 186.

Question #33. Both components rely on a layered protection system that includes a fire-resistive outer shell, vapor barrier, and thermal barrier. NFPA 1001: 5.1.1.2. *FFHB, 2E:* FFHB: Page 128.

Question #34. Clipping of a message canoccur in the beginning and end of a message. NFPA 1001: 5.2.3. *FFHB, 2E:* FFHB: Page 65-67.

Question #35. In general, anything that has a closed handle can be hoisted with a figure-eight or bowline, while longer cylindrical tools (i.e., ax and pike pole) can be hoisted using a clove hitch and half hitches. NFPA 1001: 5.1.1.1. *FFHB, 2E:* FFHB: Page 449.

Question #36. Pre-action systems are used in areas where the materials protected are of high value and water damage would be expensive, such as computer rooms and historical items. NFPA 1001: 6.5.1. *FFHB, 2E:* FFHB: Page 321.

Question #37. Caution All forcible entry operations must be coordinated with fire attack and ventilation. Lack of coordination may result in rapid fire spread or a backdraft. NFPA 1001: 5.3.4; 6.3.2. *FFHB, 2E:* FFHB: Page 531.

Question #38. Firefighter O'Malley is preparing for an upcoming presentation on fire safety to a high school class. Firefighter O'Malley should include current information in the presentation and should try to relate the subject matter to the audience. The presentation should include some "hands-on" activities, if possible, to reinforce learned concepts. NFPA 1001: 5.5.2. *FFHB, 2E:* FFHB: Page 680.

Question #39. Fire station tours should be tailored to fit the needs and interests of the visitors. The preschool children would be most receptive to a discussion on how to call for the fire department, as discussions about response procedures, detector placement, and first aid are beyond their comprehension level. NFPA 1001: 5.5.2. *FFHB, 2E:* FFHB: Page 684.

Question #40. EP=NP+FL±E+SA NFPA 1001: 5.3 NFPA 6.3. *FFHB, 2E:* FFHB: Page 92.

Question #41. Dynamic influences such as dirty gear, moisture (including perspiration), and fabric compression (from an SCBA) can reduce the TPP of the clothing. NFPA 1001: 5.1.1.2. *FFHB, 2E:* FFHB: Page 128.

Question #42. First responders should have a basic awareness of the threat of terrorism and basic response actions. NFPA 472: 4.2.1. *FFHB, 2E:* FFHB: Page 929.

Question #43. Observant firefighters on scene can provide fire investigators with reliable information to where the fire was, what spectators said about the fire, and how the fire reacted during fire attack. NFPA 1001: 6.3.4. *FFHB, 2E:* FFHB: Page 651-654.

Question #44. Pre-action systems are used in areas where the materials protected are of high value and water damage would be expensive, such as computer rooms and historical items. NFPA 1001: 6.5.1. *FFHB, 2E:* FFHB: Page 321.

Question #45. However, picking the largest fire extinguisher available to put out a small fire can make it an expensive fire. NFPA 1001: 5.3.16. *FFHB, 2E:* FFHB: Page 195.

Question #46. Solid stream handlines can reach more than 70 feet and master streams about 100 feet. NFPA 1001: 5.3 NFPA 6.3. *FFHB, 2E:* FFHB: Page 282.

Question #47. Fire departments must establish an accountability system to track personnel entering an IDLH atmosphere. Firefighters must work in teams of two as a minimum. PASS devices must be activated. NFPA 1001: 5.3.1. *FFHB, 2E:* FFHB: Page 168.

Question #48. Ventilate as you advance, as long as it will not spread the fire. Ventilation allows the products of combustion to escape and provides a better environment. Check for outside openings such as windows and doors. This will provide a means of escape in an emergency and provide the firefighters' location to outside personnel. NFPA 1001: 5.3.1. *FFHB, 2E:* FFHB: Page 150.

Question #49. •Immediately exit the hazardous environment. All team members must exit. Never leave a firefighter alone. •If necessary, use the protective hood as a filter. Use the "buddy breathing" attachment on an SCBA as a last resort because this will deplete the air supply at least twice as fast. NFPA 1001: 5.3.1. *FFHB, 2E:* FFHB: Page 171.

Question #50. Testing should be conducted on fire hydrants periodically to ensure that they are operable and to determine the flow rate of the hydrants. NFPA 1001: 5.3.15. *FFHB, 2E:* FFHB: Page 213.

Question #51. Open-circuit supplied air respirators (SARs), also called airline respirators, are similar to SCBA units, except that the air supply cylinder is remote from the user. Air is supplied in the same manner as for a regular SCBA, but the hose connecting the cylinder and the SCBA unit may be 100 to 200 feet long. This type of unit must be equipped with an SCBA escape unit with duration of approximately 5 to 10 minutes. NFPA 1001: 5.3.1. *FFHB, 2E:* FFHB: Page 158.

Question #52. Ropes constructed from polypropylene are primarily used for water rescue operations by the fire service. This is due to the fact that water has no effect on their strength, and even more beneficial is the fact that they will float. NFPA 1001: 5.1.1.1. *FFHB, 2E:* FFHB: Page 421.

Question #53. If it does freeze, it should be carefully moved to a place to thaw prior to any folding or bending. NFPA 1001: 5.3. *FFHB, 2E:* FFHB: Page 223.

Question #54. Washing structural or wildland gear along with linens or other household items should be forbidden because cross-contamination can result. NFPA 1001: 5.1.1.2. *FFHB, 2E:* FFHB: Page 136.

Question #55. It is a fact of nature that fire runs uphill faster than downhill. Caution:Weather is one of the main causes of firefighter fatalities on wildland fires. NFPA 1001: 5.3.19. *FFHB, 2E:* FFHB: Pages 595 and 598.

Question #56. The firefighter must determine which way the door swings. The common way of describing this is in relationship to the forcible entry team. Doors with exposed or visible hinges will swing toward the forcible entry team. NFPA 1001: 5.3.4; 6.3.2. *FFHB, 2E:* FFHB: Page 523.

Question #57. Instead, the firefighter will find high voltage warning labels on components. NFPA 1001: 5.3.7. *FFHB, 2E:* FFHB: Page 600.

Question #58. This method, if used indoors, will usually greatly disturb the thermal balance, cause a loss of visibility in the structure as heat and products of combustion are circulated downward. When using this method, the firefighting team should attempt to shield themselves from the steam vapor by backing out of the area after applying the water. They can reenter as soon as the vapor begins to dissipate. NFPA 1001: 5.3.10. *FFHB, 2E:* FFHB: Page 604.

Question #59. Thermal layering is an important concept for firefighters because it affects how to enter and function in a room or area that is on fire or where the fire has just been extinguished: firefighters must stay low to the floor in attacking structural fires. NFPA 1001: 5.3; 6.3. *FFHB, 2E:* FFHB: Page 99.

Question #60. Spanner wrenches come in several sizes and are used to tighten or loosen couplings. NFPA 1001: 5.1.1.1. *FFHB, 2E:* FFHB: Page 227.

Question #61. Check for proper seal by attaching the regulator and inhaling a breath, activating the regulator.Hold that breath for about five seconds and listen and feel for any air leaks. NFPA 1001: 5.3.1. *FFHB, 2E:* FFHB:Page 166.

Question #62. Keeping up with changing technology is important for overall safety and efficiency. NFPA 472: 4.2.1. *FFHB, 2E:* FFHB:Page 743.

Question #63. The ideal balance of protein, fat, and carbohydrates to stabilize the release of insulin into the bloodstream is: 30% protein, 30% fat, 40% carbohydrates. NFPA 1001: 5.3; 6.3. *FFHB, 2E:* FFHB: Page 732.

Question #64. The point here is to avoid standing in front of any bumper after it has been heated. NFPA 1001: 5.3.7. *FFHB, 2E:* FFHB: Page 600.

Question #65. Substituting other liquids for water can slow the absorption of water into the system. For that reason, just water should be given for the first hour. For activities lasting longer than an hour, some consideration can be given to adding essential electrolytes and nutrients along with the water. NFPA 1001: 5.3; 6.3. *FFHB, 2E:* FFHB: Page 732.

Question #66. Compare the cylinder and regulator gauges. Gauge readings should be within 100 psi. NFPA 1001: 5.3.1. *FFHB, 2E:* FFHB: Page 172.

Question #67. This customizing should only be performed by a certified manufacturer that understands and complies with NFPA standards for both structural PPE and life safety system components. NFPA 1001: 5.1.1.2. *FFHB, 2E:* FFHB: Page 128.

Question #68. The biggest problem with PASS devices results when wearers simply forget to turn their units on. This simple mental lapse has contributed to numerous firefighter fatalities. NFPA 1001: 5.1.1.2. *FFHB, 2E:* FFHB: Page 135.

Question #69. A technique that has produced satisfactory results when operating off a flat roof beyond the reach of a ladder is to use a tool and a rope. NFPA 1001: 5.3.11. *FFHB, 2E:* FFHB: Page 570.

Question #70. Dry chemical or dry powder extinguishers carried on apparatus should occasionally be rotated upside down and shaken to keep powders from packing at the bottom as a result of vehicle vibration. NFPA 1001: 5.3.16. *FFHB, 2E:* FFHB: Page 197.

Question #71. A group is a functional designation to conduct a specific task, such as search and rescue or ventilation. NFPA 1001: 6.1.1.1. *FFHB, 2E:* FFHB: Page 42.

Question #72. Several firefighters are discussing the growth stage of fire. The speed of the growth and ultimately the size of the fire are dependent on oxygen supply, fuel, container size, and insulation. NFPA 1001: 5.3; 6.3. *FFHB, 2E:* FFHB: Page 92.

Question #73. The logistics section chief is responsible for securing the facilities, services, equipment, and materials for an incident. NFPA 1001: 6.1.1.1. *FFHB, 2E:* FFHB: Page 40.

Question #74. Ladders should be stored with multiple support points to prevent bowing or sagging. NFPA 1001: 5.3.6; 5.3.9; 5.3.10; 5.3.11; 5.5.3. *FFHB, 2E:* FFHB: Page 381.

Question #75. This may require using the floor below and advancing the hoseline toward the fire location. NFPA 1001: 6.5.1. *FFHB, 2E:* FFHB: Page 334.

Question #76. Great success can be achieved with positive pressure ventilation or with the use of a hoseline stream to create negative pressure and pull the smoke out. NFPA 1001: 5.3.11. *FFHB, 2E:* FFHB: Page 588.

Question #77. Firefighters must remember to have a path of escape behind them and not allow the fire to get between them and the exit. NFPA 1001: 5.3.16. *FFHB, 2E:* FFHB: Page 195.

Question #78. Firefighter Lopez is inspecting an extension ladder. The halyard should be inspected for undue wear. The heat sensor label should be checked for presence and condition. The ladder should be examined for any discoloration from heat damage, cracks, and deformed areas. NFPA 1001: 5.3.6; 5.3.9; 5.3.10; 5.3.11; 5.5.3. *FFHB, 2E:* FFHB: Page 382.

Question #79. In the 1970's, the US government commissioned a panel of fire service experts to study the country's changing fire problem. This group was called the National Commission on Fire Protection and Control. As a result of the document this group published, America Burning, many positive changes occurred in fire service at various levels of government, including, at the federal level, the creation of the United State Fire Administration and the National Fire Academy. NFPA 1001: N/A. *FFHB, 2E:* FFHB: Page 18.

Question #80. PPE provides a minimum level of protection and should be considered the last resort of protection for firefighters and emergency responders operating at an incident. NFPA 1001: 5.1.1.2. *FFHB, 2E:* FFHB: Page 125.

Question #81. This system allows the pump to operate at full flow, but may cause damage to the pump and valves without proper back-flushing of the entire pump. NFPA 1001: 5.3 NFPA 6.3. *FFHB, 2E:* FFHB: Page 298.

Question #82. Understanding toxicology and health effects is important. NFPA 472: 5.2.2; 5.2.3. *FFHB, 2E:* FFHB: Page 838.

Question #83. Common overhaul tools are pike poles, pitchforks, rubbish hooks, shovels, axes, chain saws, carry-alls, and wheelbarrows. NFPA 1001: 5.3.13. *FFHB, 2E:* FFHB: Page 649.

Question #84. Loosen the straps and rotate the SCBA unit under the arm along the rib cage as a first step. NFPA 1001: 5.3.1. *FFHB, 2E:* FFHB: Page 170.

Question #85. NFPA requires manufacturers to clearly label care instructions for cleaning each piece of equipment. In addition, manufacturers should provide the user with specific instructions and information that addresses the limitations of use. NFPA 1001: 5.1.1.2. *FFHB, 2E:* FFHB: Page 136.

Question #86. NFPA 1500 covers the following topics: Administration Training and education Vehicles, equipment, and drivers Protective clothing and equipment Emergency operations Facility safety Medical and physical Member assistance and wellness Critical Incident Stress Program NFPA 1001: 6.1.1.1. *FFHB, 2E:* FFHB: Page 109.

Question #87. "Black fire" is apparent when a team of firefighters prepares to initiate fire attack. The smoke is high volume with a turbulent velocity and is ultradense and black. They should expect autoignition or flashover when they "read" this smoke condition. NFPA 1001: 5.3; 6.3. *FFHB, 2E:* FFHB: Page 103.

Question #88. As security requirements increase, home and business owners have begun to install many varieties of locks and security devices. If these types of devices are in use, the forcible entry team may need to find an alternate means of entry or use a rotary saw to gain access. NFPA 1001: 5.3.4; 6.3.2. *FFHB, 2E:* FFHB: Page 530.

Question #89. The signs and symptoms of internal bleeding are pale skin; cold and clammy skin; bruising of the skin; dilated pupils; obvious deformities to major bones; a rigid and tender abdomen; blood in the urine or from the rectum; and blood from the mouth, nose, or blood in vomitus. NFPA 1001: 4.3. *FFHB, 2E:* FFHB: Page 709-710.

Question #90. The follow-through figure-eight knot is very useful when attaching a utility or life-safety line rope to an object that does not have a free end available. NFPA 1001: 5.1.1.1. *FFHB, 2E:* FFHB: Page 433.

Question #91. The positive-pressure technique actually injects air into the compartment and pressurizes it. In an attempt to equalize, the smoke and heat are carried out into the areas of lower pressure outside the structure. NFPA 1001: 5.3.11. *FFHB, 2E:* FFHB: Page 573.

Question #92. To disconnect the opener, break out a panel near the attachment mechanism, reach in with a tool to grab the release cord, and pull. NFPA 1001: 5.3.4; 6.3.2. *FFHB, 2E:* FFHB: Page 525.

Question #93. The straight or storage hose roll is the easiest with which to work. Start with the hose flat on the ground. From the male end, to protect the threads, roll straight to the opposite end. NFPA 1001: 5.1.1.1. *FFHB, 2E:* FFHB: Page 231.

Question #94. Firefighters arrive on a scene to find a Class C fire. The best method of attack is to remove the flow of electricity and then re-analyze the class of the fire for further actions. A Class A or B fire may be left after the Class C fire is resolved. NFPA 1001: 5.3; 6.3. *FFHB, 2E:* FFHB: Page 100-101.

Question #95. Pressurized flammable liquids and gases are special fire hazards that should not be extinguished unless the fuel can be immediately shut off. NFPA 1001: 5.3.16. *FFHB, 2E:* FFHB: Page 186.

Question #96. When Firefighter Hernandez is ready to transmit his message, he should first listen to the radio to make sure he does not interfere with other communications. When the airwaves are clear, he should depress the push-to-talk button and wait at least two seconds before beginning to speak to avoid clipping his message. NFPA 1001: 5.2.3. *FFHB, 2E:* FFHB: Page 65-67.

Question #97. Streetsmart Tip The rescuer must remember that none of these drags provides spinal immobilization and are intended to be utilized only in situations where greater harm will come to the patient if not immediately moved. NFPA 1001: 5.3.9. *FFHB, 2E:* FFHB: Page 477.

Question #98. Firefighters can look for obvious signs of hidden fire and use their senses by feeling the wall for heat, smelling around outlets and other openings for a scent of anything burning, and listening for sounds of items burning. NFPA 1001: 5.3.13. *FFHB, 2E:* FFHB: Page 649.

Question #99. Firefighters arrive on a scene to find a Class A fire. The best method of attack is to cool the fire with water. NFPA 1001: 5.3; 6.3. *FFHB, 2E:* FFHB: Page 100.

Question #100. Steel has excellent tensile, shear, and compressive strength. NFPA 1001: 5.3.10; 6.3.2. *FFHB, 2E:* FFHB: Page 349.

PHASE IV

FINAL EXAM

This the final section in the Exam Prep Guide. For this section, we addressed all levels of Bloom's Taxonomy, Cognitive Domain, and all the previous sections. When taking the Final Exam from this section, you will find a variety of questions from basic understanding to application level questions. One should have successfully completed the previous sections before attempting Section Four. Successful completion of this section would indicate a strong knowledge of the material and an in-depth understanding of the content.

1. You have responded to a commercial fire alarm in a local hardware/building supplies chain.The 2 A.M. alarm has used all available fire resources as there is another fire alarm in an apartment complex across town. The number one priority on both of these fire scenes is rescue/evacuation.

    a. True

    b. False

2. An engine company arrives on the scene of a large storage building that is fully involved in flame with heavy smoke conditions. Bystanders in the next building report that the shed on fire is unoccupied and very little remained in storage in the facility. Using the concept of risk/benefit in this situation, you can assume that the team will make entry into the structure for search and rescue and an aggressive fire attack.

    a. True

    b. False

3. Setting a knot is the practice of making sure all parts of the knot are lying in the properorientation to the other parts and look exactly as the pictures indicated.

    a. True

    b. False

4. While enjoying an off-duty afternoon in a movie theater, you find a seat burning in the second row. You tell your significant other to call 9-1-1 while you search for an extinguisher. You find four extinguishers with the older symbols on them. You know you want to pick the extinguisher with a yellow star and a letter inside it.

    a. True

    b. False

5. Protective systems have a major drawback in that they require a human to detect the fire and manually activate an alarm.

    a. True

    b. False

6. You are advancing toward a very large fire in a commercial business. You should ventilate by taking out the windows as you advance.

    a. True

    b. False

7. As you advance up the stairwell of a multistory residential building with a fire alarm, you know that you will need hose to connect to the standpipe. You can use either hose from your engine or the hose stored in the cabinet.

    a. True

    b. False

8. Every communications facility should be supported by a backup location in the event that the primary facility encounters problems that render it inoperable and result in evacuation.

    a. True

    b. False

9. A bowline knot is the only knot that is recommended for tying webbing.
    a. True
    b. False

10. It is imperative for all firefighters to be familiar with some sort of rescue knot.
    a. True
    b. False

11. You are in staging and watching the scene of a structure fire. You see one of the attack lines suddenly jump off the ground. You can suspect a water hammer.
    a. True
    b. False

12. Emergency communications centers should be equipped with backup power supplies.
    a. True
    b. False

13. Firefighting is a team effort. Working alone or outside the action plan endangers individuals and the team.
    a. True
    b. False

14. It is more effective for firefighters to determine the type of construction and fire-resistive rating once a building is complete as opposed to viewing the building during construction.
    a. True
    b. False

15. The call-taking process consistes of receiving a report, intervening, and referral or dispatch composition.
    a. True
    b. False

16. Hoselines can be hoisted using ropes, but they must be uncharged.
    a. True
    b. False

17. Because regular training is the single most important step in firefighter safety, the firefighter must strive to retain the information and skills that are presented in training sessions.
    a. True
    b. False

18. A firefighter has raised a ladder to a structure and is determining the proper distance the foot or butt of the ladder should be away from the building. The firefighter can estimate this by moving the foot of the ladder away from the building a distance of one-half the working length of the ladder.
    a. True
    b. False

19. First responders should always follow the "use water to knock down vapors" recommendation in the DOT ERG.
    a. True
    b. False

20. When connecting to a standpipe inside a structure, you must have faith that the driver/operator will complete his portion of the evolution.
    a. True
    b. False

21. When selecting a fire extinguisher, it is very important that a firefighter be able to classify the fire into one of 5 classes.
    a. True
    b. False

22. A solid tip is a type of fog nozzle.
    a. True
    b. False

23. Out-of-service hose can be put back into service only if it is tested on its annual test due date.
    a. True
    b. False

24. A material that is heavy and dense will be a less effective thermal conductor than a material that is light and less dense.
    a. True
    b. False

25. Incident readiness for firefighters includes personal protective equipment, personal accountability, and fit-for-duty status.
    a. True
    b. False

26. Which of the following is not true concerning school evacuation drills?
    a. Specific exits should be assigned to groups of classrooms and all exits should be used to provide for even distribution of exiting occupants.
    b. Emergency evacuation plans should be drawn in graphic form and posted at each exit.
    c. The focus of the drill should be disciplined control and order. Speed will result from properly planned and supervised evacuations.
    d. The fire department should be present whenever possible at evacuation drills.

27. The four elements of a fire stream are _____.
    a. pump, water, hose, and nozzle
    b. source, hose, pressure, and nozzle
    c. psi, gpm, length of hose, and friction loss
    d. type of source, diameter of hose, length of hose, and gallons per minute

28. The pressure in the system with no hydrants or water flowing is _____ pressure.

    a. residual

    b. static

    c. atmospheric

    d. flow

29. Which category of WMD agents can only be treated by moving the victims to fresh air?

    a. Blister agents

    b. Nerve agents

    c. Nuclear

    d. Irritants

30. You and your partner are conducting a search operation in a smoke-filled residential structure. You find two unconscious men in the basement near a pool table. The men are both of equal size and appear to weigh in excess of 200 lbs. Your partner weighs 118 lbs and does not have a lot of upper body strength. You suggest she use a _____ to move the second man to the bottom of the stairs.

    a. foot drag

    b. a webbing sling drag

    c. firefighter's carry

    d. seat carry

31. A group of firefighters are discussing the storage of ladders. Which of the following is incorrect about ladder storage?

    a. Ladders should be stored in clean, dry places.

    b. Ladders should be stored with one support point only.

    c. Ladders should be placed in a storage area where they will not be subjected to heat or exhaust.

    d. Ladders should not be placed in out-of-sight areas where they can be a tripping hazard.

32. Your engine has parked at the scene of a structure fire. The driver operator tells you she is going to use a reverse lay. She is referring to _____.

    a. a direct attack

    b. an indirect attack

    c. a master stream

    d. water supply

33. _____ hose is used by trained personnel to fight fires.

    a. Attack

    b. Booster

    c. Supply

    d. Soft suction

34. Seat-mounted SCBA _____.

    a. allows a firefighter to don the SCBA en route to an emergency

    b. allows for less cylinder failures due to the storage technique

    c. interferes with bunker gear donning en route to a call

    d. can interfere with the cylinder gauge check

35. The permissible exposure limit is an average of _____ for a chemical exposure.

    a. 15 minutes

    b. 8 hours

    c. 24 hours

    d. 1 hour

36. Fire hose is made in three types of construction. Which of these is not one of these types?

    a. Wrapped

    b. Plastic

    c. Braided

    d. Woven

37. You have been working the interior of a residential structure fire and find yourself low on air. You and your partner retreat through the garage and you find the door will not raise. The power has been disconnected to the house so the automatic garage door opener will not work. The easiest method of exit is to _____.

    a. take out a window on the door to create an exit

    b. take out a panel of the door with your axe

    c. make a door in the garage wall

    d. pull the release rope

38. As a conscientious firefighter, you realize that dynamic influences can reduce the TPP of your structural gear. Fabric compression is one of these factors resulting from _____.

    a. folding your structural pants and placing them under your helmet while the gear is in storage

    b. wearing webbing that carries an axe and a flashlight

    c. wearing an SCBA

    d. all of the above

39. Firefighter Morton is giving instructions to the operator of an aerial ladder. She wants to give the command to move the ladder from a horizontal position to a vertical angle. What command should she use?

    a. Raise

    b. Extend

    c. Retract

    d. Rotate

40. What incident command designation is established to maintain span of control over a number of divisions, sectors, and/or groups?

    a. A strike team

    b. A branch

    c. A section

    d. A task force

41. Your engine is attached to the hydrant using two 2½" supply lines. You notice that one of the lines is leaking a considerable amount of water at one of the couplings. You should _____.

    a. pull another section of 2½" line for replacement

    b. tighten the coupling with a hydrant wrench

    c. tighten the coupling with spanner wrenches

    d. use a hose bridge

42. Firefighter Jong is searching for a new federal OSHA regulation about respiratory protection. Where would she look to find this federal regulation?

    a. United States Fire Administration manuals

    b. The Code of Federal Regulations

    c. National Fire Protection Association standards

    d. The National Institute for Occupational Safety and Health reports

43. A(n) _____ is a scrape or brush of the skin usually making it reddish in color and resulting in minor capillary bleeding.

    a. avulsion

    b. abrasion

    c. incision

    d. laceration

44. A standard wildland hose load is the modified _____.

    a. Gasner bar pack

    b. horseshoe pack

    c. reverse-drain and carry pack

    d. shoulder loop

45. The progressive end of a ground cover or wildland fire is known as the _____.

    a. head

    b. tail

    c. arrow

    d. flank

46. While bagging life-safety rope, you remind your partner that he should _____ the rope to prevent hang-up.

    a. stuff

    b. coil

    c. wad

    d. fold

47. Nozzle reaction is _____.

    a. the movement of the nozzle opposite from the direction of the water flow

    b. the movement of the nozzle in the same direction as the water flow

    c. dangerous to persons weighing less than the nozzle operator

    d. inversely proportionate to the gpm

48. Within an Incident Management System, there are four section chiefs: operations, logistics, _____, and finance/administration.

    a. safety

    b. staging

    c. planning

    d. command

49. You are an A shift firefighter. At morning pass down, B shift engine company neglects to inform you they used a pressurized-water extinguisher on a mattress fire. You discover this when you go to use the extinguisher on a small trash fire in an office building. This scenario tells you something important. What is it?

    a. B shift company should be written up.

    b. A shift company officer is neglectful.

    c. B shift company officer is neglectful.

    d. A shift morning truck checks were not thoroughly completed.

50. A group of preschool children have come to visit the fire station. Firefighter Williams is giving a tour of the facility. Which of the following things should be focused on during his presentation to the preschool children?

    a. How to call for the fire department in an emergency

    b. Fire department response procedures

    c. Proper placement of carbon monoxide detectors

    d. Upcoming first aid classes

51. A rehabilitation area at the fireground is a _____ effort to prevent injury.

    a. reactive

    b. proactive

    c. both a and b

    d. none of the above

52. As a tender shuttle driver/operator, you should pick a dump site that has

    _____.

    a. access

    b. turnaround area

    c. safety

    d. all of the above

53. Installed fire-extinguishing systems _____.

    a. allow SCBA use to be discontinued

    b. can create an oxygen-deficient atmosphere

    c. extinguish the fire and do not require firefighters to fight fire offensively

    d. require SCBA to have a low-pressure breathing hose

54. You are searching for the seat of a fire with your partner in the interior of a residential structure fire. As you advance down a hallway, you find you are following an exterior wall with windows and a door. The products of combustion are thick and visibility is just inches. However, you do not feel an increase in ambient temperature. Your course of action should be to _____.

    a. retreat and find other access to the fire

    b. notify IC and retreat to find other access to the fire

    c. break out windows as you advance

    d. hold your position at the mouth of the hallway and have your partner advance toward the fire

55. During pure combustion, energy is released from an exothermic reaction as heat and

    _____.

    a. light

    b. smoke

    c. carbon

    d. carbon monoxide

56. A _____ is used to secure the firefighter to the ladder when both hands must be used to perform a task and a ladder belt is not available.

    a. rung lock

    b. halyard

    c. leg lock

    d. carry lock

57. _____ is the measurable amount of pressure being exerted against a confined container by a liquid substance as it converts to a gas.

    a. Vapor density

    b. Evaporation

    c. Vapor pressure

    d. Diffusion

58. PPV uses a technique that _____.

    a. sucks smoke

    b. pressurizes smoke

    c. replaces smoke with water

    d. removes smoke with water

59. You are responding to an MVA with trapped occupants and extrication required. It is a hot July day. It is 1 p.m. and you have just returned from fighting a large dumpster fire. The temperature outside is 100°F and it is getting hotter. Which is the appropriate course of action concerning PPE?

    a. Wait until arrival on scene, then ascertain the proper amount of PPE to don, leaving off what is deemed unnecessary.

    b. Don full PPE even though your liners are a little moist. Leave jacket and trousers unfastened to allow ventilation.

    c. Pull the liners out of your PPE and don the outer shell and SCBA.

    d. Don full PPE and drink water en route to the scene. Dress down on scene, if allowed and appropriate.

60. Which of the following is a commonly used tool in salvage operations?

    a. Pike pole

    b. Water vacuum.

    c. Rubbish hook

    d. Axe

61. OSHA requires two important positions be assigned at a hazardous materials incident. One is the incident commander. What is the other position?

    a. Remediation officer

    b. Entry officer

    c. Degradation officer

    d. Safety officer

62. Reports of emergencies can be received in a variety of ways. Which of the following is a way in which a report of emergency might be received?

    a. Emergency call boxes

    b. Wireless or cellular phones

    c. Automatic alarms

    d. All of the above

63. Firefighters are responding to a fire in a structure that is Type V construction. The most common occupancy of this type of construction is _____.

    a. residential

    b. commercial

    c. business

    d. educational

64. Many units from the Delmar Fire Department are on the scene of a fire in an apartment building. The incident commander wants to check on the availability of getting additional hose from the fire department storage area to use on the scene. Who should he contact?
    a. The operations section chief
    b. The planning section chief
    c. The logistics section chief
    d. The finance/administration section chief

65. Two firefighters are raising an extension ladder to a structure. This ladder will be used by many firefighters to access the structure. Where can the firefighters find the maximum load capacity of the ladder in order not to overload it?
    a. On the manufacturer's label affixed to the ladder
    b. Engraved in the bottom rung of the ladder
    c. Engraved in the top rung of the ladder
    d. It is not located on the ladder. The firefighters have to remember the information.

66. The accident chain includes the following components: the environment, _____, equipment, the event, and the injury.
    a. personnel policies
    b. procedures
    c. human factors
    d. mitigation

67. You are on a rescue operation of a single-family residence with an unconscious person inside. There are security bars and gates on all windows and doors. There has been a flash flood and the home now has about one foot of water in it. Because of the layout and the terrain, the garage is still dry, but the water is rising. You have just used a rotary saw to cut into a garage door. You find that the door into the home from the garage is secured with a padlock on the garage side of the door. Your partner is carrying a flashlight and has no other tools. You should _____.
    a. use the saw to cut the padlock, one shackle at a time
    b. use the saw to cut the padlock, both shackles at the same time
    c. radio command for an alternate means to cut the lock
    d. return to staging

68. You have just finished extinguishing a car fire with a 2½" line. The car was a total loss. The temperature outside is 90°F. You are attempting to open the hood. There is a small amount of steam and/or smoke coming from the front wheel wells. At this point in the fire, you may _____.
    a. remove your SCBA as long as you stay 5 feet from the vehicle
    b. remove your SCBA and coat, even in the vicinity of the vehicle as the hot temperature rules now apply
    c. remove your helmet to allow heat to leave your body as long as you are 5 or more feet from the vehicle
    d. stay in full PPE as long as you are near the vehicle and there is possibly products of combustion

69. The ACGIH recommends that you only be exposed to 20 ppm of a chemical; while OSHA requires that the exposure be less than 35 ppm. To which one do you have to be legally adhered?

    a. OSHA

    b. ACGIH

    c. NIOSH

    d. None of the above

70. Foam usually uses _____ to create a blanket over the surface of the fuel to cool and smother the fire, while sealing the escaping vapors.

    a. air

    b. carbon monoxide

    c. O2

    d. argon

71. Once you don your face piece, you should check for a proper seal by _____.

    a. placing your hand over the regulator hole and attempting to inhale

    b. attaching the regulator, taking a breath and holding it, and listening and feeling for any air leaks

    c. smelling for smoke or fumes once you enter the IDLH

    d. No check is needed. It is mandated by NFPA that all masks be custom fit.

72. NFPA 1581 requires that clothing be cleaned _____ as a minimum.

    a. once a week

    b. once a month

    c. yearly

    d. every six months

73. You respond to a reported smoke detector activation in the diorama room of a museum. Upon arrival, you are told there is a sprinkler system, but it has not activated. Based on this information, you can be relatively sure that the system is a _____ system.

    a. deluge

    b. dry pipe

    c. wet pipe

    d. pre-action

74. Stored hose _____.

    a. can last indefinitely

    b. should be used occasionally

    c. should occasionally be refolded

    d. can last for two years and then must be discarded

75. You are fighting a fire in a commercial warehouse membership club. The fire is in one of the stacks of fertilizer bags. While it is burning, the immediate concern is _____.

    a. hazmat

    b. exposure

    c. collapse

    d. smoke

76. SCBA must be maintained _____.

    a. daily

    b. monthly

    c. annually

    d. all of these

77. As a firefighter, you know that anytime you encounter a PIV, WIV, or OS&Y, it should be _____.

    a. in the open position

    b. in the closed position

    c. locked in the open position

    d. locked in the closed position

78. Firefighters should never enter a _____ environment without first engaging SCBA.

    a. total flooding

    b. deluge

    c. wet pipe

    d. none of the above

79. _____ are supervisory-level positions and are responsible for both firefighters and administrative duties.

    a. Firefighters

    b. Rescue specialists

    c. Company officers

    d. Driver/operators

80. _____ is the process of minimizing the chance, degree, or probability of damage, loss, or injury.

    a. Risk management

    b. The safety chain

    c. The safety triad

    d. Vicarious experience

81. You are in mop-up mode on a structure fire where you used a considerable amount of foam. The driver/operator of the engine tells you the foam system on the apparatus uses an around-the-pump proportioner. You should expect to _____.

    a. deplete the entire foam capacity

    b. back-flush the pump

    c. prime the pump

    d. discard the first section of hose

82. Your engine is attached to the hydrant using two 2½" supply lines. You notice that one of the lines is leaking a considerable amount of water at one of the couplings. You should _____.

    a. pull another section of 2½" line for replacement

    b. tighten the coupling with a hydrant wrench

    c. tighten the coupling with spanner wrenches

    d. use a hose bridge

83. Pre-piped water lines for firefighter fire streams in a building are known as _____ systems.

    a. sprinkler

    b. standpipe

    c. hydrant

    d. exterior water supply

84. During fire safety inspections, firefighters should check fire sprinkler systems to ensure that all water supply vales are _____ and secured.

    a. closed

    b. hidden

    c. open

    d. capped

85. You are driving a rescue truck to the scene of a motor vehicle accident on a busy highway. Your first priority, upon arrival, should be to _____.

    a. position forward of the scene 100 feet

    b. position behind the scene 100 feet

    c. access the scene and establish a work area

    d. position so as not to impede traffic flow

86. An important agent maintenance duty for dry chemical extinguishers is to _____.

    a. wash the exterior with soap and water

    b. visually examine the exterior for soap residue

    c. inspect the bracket

    d. turn them upside down periodically

87. You are explaining hydraulics to your rookie firefighters. You explain the formula for calculating engine pressure. One of the students does not understand the formula as you explain it. Therefore, you write it on the board as _____.
    a. EP=NP-FL±E+SA
    b. EP=NP*FL±E+SA
    c. EP=NP/FL±E+SA
    d. EP=NP+FL±E+SA

88. You are on a rescue scene. You are attempting to extend a rope for a rescue operation, as it is too short. You know that you have a duplicate rope in your apparatus. You know that you should consider the _____ to assist your operation.
    a. Becket bend
    b. bowline
    c. water knot
    d. figure-eight

89. To determine how many additional lines can be connected to a fireground operation, it is vital that you remember to check the _____ pressure before any lines are opened.
    a. static
    b. residual
    c. flow
    d. nozzle

90. You are on a rescue scene. You are attempting to extend a rope for a rescue operation, as it is too short. You know that you have a duplicate rope in your apparatus. You know that you should consider the _____ to assist your operation.
    a. Becket bend
    b. bowline
    c. water knot
    d. figure-eight

91. "Black fire" is apparent when a team of firefighters prepares to initiate fire attack. The smoke is high volume with a turbulent velocity and is ultradense and black. What should the firefighters expect to occur when they "read" this smoke condition?
    a. Backdraft
    b. Flash point
    c. Autoignition and flashover
    d. The decay phase

92. You are securing your life-safety rope around an anchor and are preparing to rappel down a 10-foot drop off. There are no trees or other suitable anchors in the area. You decide to use the axle of your apparatus. You use which knot to accomplish this anchor?
    a. figure-eight follow-through
    b. Becket bend
    c. clove hitch
    d. half hitch

93. A 4A rating of an extinguisher should extinguish _____.

    a. 4 square-feet of fire

    b. 4 times the fire of a 1A

    c. 16-square-feet of fire

    d. 4 lbs of class A material

94. A(n) _____ is a catalyst in the breakdown of molecules and possesses a chemical property that can pull apart a molecule and break apart the bond that previously existed.

    a. oxidizer

    b. proton

    c. electron

    d. hydrocarbon

95. _____ occurs when the entire contents of a compartment ignite almost simultaneously, generating intense heat and flames.

    a. Thermal layering

    b. Rollover

    c. Flashover

    d. Backdraft

96. You are a firefighter assigned to the airport station. You respond to a reported in-flight emergency that might result in a fiery crash. En route to the staging point, you realize you have your structural helmet instead of your proximity helmet. Your concern would be _____.

    a. radiant facial burns from a melting shield

    b. your helmet is black and not silver

    c. less impact protection from special grade aircraft structural members

    d. "ARFF blindness" due to special coatings not being applied to your structural face shield

97. A team of firefighters exits a structure after completing their assigned task. The team is told to report to the _____ area for rest and hydration.

    a. staging

    b. accountability

    c. rehabilitation

    d. command post

98. You are explaining hydraulics to your rookie firefighters. You explain the formula for calculating engine pressure. One of the students does not understand the formula as you explain it. Therefore, you write it on the board as _____.

    a. EP=NP-FL±E+SA

    b. EP=NP*FL±E+SA

    c. EP=NP/FL±E+SA

    d. EP=NP+FL±E+SA

99. On the scene of a commercial structure fire, you are connecting hose to the Siamese FDC. You always connect to the left outlet first _____.
    a. to allow better access for using the spanner wrench to tighten the coupling
    b. as some outlets only have clappers on the right side
    c. both of these
    d. neither of these

100. When responding to residential structure fires with sprinkler systems, it is important for firefighters to remember that these systems _____.
    a. are intended to protect vital areas, such as computer rooms or stereo centers
    b. are only intended to protect egress hallways
    c. are intended for life safety and not necessarily property conservation
    d. are intended for property conservation and not necessarily life safety

101. You are doing a monthly maintenance on the SCBA on your engine. You find that the shoulder harness has a half-inch cut in it. You should _____.
    a. log the information on the monthly maintenance sheet
    b. burn the edges of the nylon to keep the cut from tearing more and place the pack back into service
    c. remove the SCBA from service until a technician can replace the damaged parts
    d. do nothing as you are not an SCBA technician

102. What is the ideal method for transfer of command?
    a. By face-to-face meeting
    b. By telephone
    c. By radio
    d. By written report

103. Although all PPE requires pre-hydration, which level is it most important?
    a. A
    b. B
    c. C
    d. D

104. You are advancing a hoseline into a commercial building which is a telemarketing center with many cubicles. As you enter, you look up and notice the ceiling tiles appear to be moving up and down. You decide to _____.
    a. retreat and regroup
    b. unseat the tiles with the hose stream
    c. advance quickly to the seat of the fire
    d. fight the fire from the doorway position

105. A SARA Title III facility should be able to provide a _____ to emergency responders.
    a. MSDS
    b. DOT ERG
    c. NIOSH pocket guide
    d. Offloading truck

106. Which of the following parts of a ladder can be found on an aerial ladder apparatus but not on a ground ladder?

    a. Beams

    b. Rails

    c. Hydraulic pistons

    d. All of the above

107. A firefighter discovers that a facility with outdoor storage of hazardous materials has a six-inch curb around the storage area. What is this called?

    a. Primary containment

    b. Primary hazard control

    c. Secondary containment

    d. Secondary hazard control

108. You are assigned to "wrap the hydrant" on the way into the fire. The engine stops, you pull off the layout load from the hose bed, and wrap the hydrant. Training has taught you that your next step is to _____.

    a. connect the hose

    b. position yourself

    c. notify the IC

    d. tell the engine to move advance

109. The public information officer arrives on the scene of a mass casualty incident where the incident command organizational structure has already been established. To whom will the public information officer report in the command structure?

    a. The operations section chief

    b. The incident commander

    c. The planning section chief

    d. The finance/administration section chief

110. When choosing PPE footwear, you should weigh the _____.

    a. cost

    b. advantages

    c. disadvantages

    d. all of the above

111. Which is not one of the key factors in the FBI's definition of terrorism?

    a. The type of explosive

    b. Intimidation

    c. Violent act

    d. Political and social objectives

112. You and your company are on scene of a shed fire at a residence. Your officer tells you to advance a line into the backyard and protect the exposures. Upon entering the backyard, you see the house has a swimming pool. On this scene, you should be most concerned about _____.

    a. accidental drowning

    b. poisonous gas

    c. children remaining in the home

    d. efficient use of water

113. Which of the following methods can be used to control external bleeding?

    a. Direct pressure on the site of bleeding

    b. Elevation of the bleeding injury

    c. Using pressure points

    d. All of the above

114. Which of the following developments in the American fire service did not occur in the period during and after the Civil War?

    a. Dalmatians became known as mascots of the fire service.

    b. Firefighters banded together to form a labor union.

    c. The use of military rank structures and command and control tactics became common in the fire service.

    d. Horses began to be used to pull the heavy steam-powered fire apparatus.

115. As an active firefighter, you should realize that PPE provides you the _____ level of protection and should be considered the _____ of protection.

    a. minimum / last resort

    b. maximum / front line

    c. minimum / front line

    d. maximum / last resort

116. _____ can be defined as a continuous mental evaluation of firefighters' immediate environments, facts, and probabilities.

    a. Freelancing

    b. Personal assignment

    c. Personal accountability

    d. Personal size-up

117. You are fighting an outdoor propane tank fire with your two partners. There is another hoseline advancing with three more firefighters and you are working in tandem. As you approach, you notice the heat appears to be increasing. You should _____.

    a. retreat

    b. decrease your fog pattern

    c. increase your fog pattern

    d. wait for the other team to advance

118. Firefighters arrive on a scene to find a Class C fire. Which of the following methods should the firefighters use to best control this type of fire?

    a. Cooling the fire with large quantities of water

    b. The application of a smothering agent to prevent oxygen from getting to the fuel

    c. Removing the flow of electricity and then re-analyzing the class of the fire for further actions

    d. Applying a dry powder to stop the chemical reaction

119. You are preparing to hoist a ladder to the third floor of a commercial building. You remember that you will use the _____ of the rope for work such as this.

    a. working end

    b. standing end

    c. standing part

    d. running end

120. You are setting up a positive pressure ventilation of a residence filled with smoke. The fire has been put out. You know that the quickest and easiest set up to prepare for overhaul is usually _____ ventilation.

    a. negative pressure

    b. positive pressure

    c. combination

    d. natural

121. When initially ventilating vertically, sometimes _____ can be initiated before a roof cut is made.

    a. opening a dumbwaiter bulkhead

    b. opening a first floor door

    c. opening a fire floor window

    d. turning on the window air conditioner

122. While on your way to a reported aircraft fire, you are going over your plan of action in your head. An advantage of the foam in your tank is that it _____.

    a. clings to vertical surfaces

    b. expands greatly

    c. does not require a special foam nozzle

    d. There are no advantages.

123. The information provided to an emergency communications center by mobile or portable radio must be _____.

    a. accurate

    b. clear

    c. complete

    d. all of the above

124. One firefighter is preparing to use a salvage cover to protect a large, unbreakable item. Which is the best deployment method to use for this situation?

    a.   Counter payoff deployment

    b.   Balloon toss

    c.   Shoulder toss

    d.   Roll toss

125. The most common type of construction for natural fiber ropes is _____.

    a.   braided

    b.   braid-on-braid

    c.   kernmantel

    d.   laid

126. A communications center may also be referred to as a PSAP, or public safety _____ point.

    a.   awareness

    b.   alarm

    c.   arrival

    d.   answering

127. While on a structure fire, you notice you are having trouble fastening your chin strap on your helmet. You decide that practice sessions of donning your helmet with gloves on at the station would be a good idea, as this will increase _____ and reduce the frustration of lost dexterity.

    a.   blood flow

    b.   muscle memory

    c.   confidence

    d.   training hours

128. The most important single forcible entry tool used in the fire service is the _____.

    a.   axe

    b.   Halligan tool

    c.   Pulaski tool

    d.   maul

129. Perhaps the most important task of using a non-integrated PASS device is ensuring _____.

    a.   it can be heard outside

    b.   it is turned on

    c.   it is equipped with extra batteries

    d.   the flashing light must be seen through smoke

130. You are responding to a reported wildland fire incident. Before donning the wildland ensemble, you should ensure _____.

   a. fire-resistive or cotton undergarments are worn

   b. a bandana is tied around your neck

   c. boots are leather with a steel toe

   d. the fire shelter is open and ready to deploy

131. In addition to voice communications, _____ 9-1-1 service provides emergency communications centers with the telephone number and address of the phone from which the call is originating.

   a. enabled

   b. basic

   c. encoded

   d. enhanced

132. Your engine driver/operator tells you she has been given the task of charging the deck gun. You know that _____.

   a. two or more firefighters will be needed to advance this line

   b. she wants you out of the way

   c. massive water is needed on this operation

   d. she is setting up for an offensive attack

133. When responding to residential structure fires with sprinkler systems, it is important for firefighters to remember that these systems _____.

   a. are intended to protect vital areas, such as computer rooms or stereo centers

   b. are only intended to protect egress hallways

   c. are intended for life safety and not necessarily property conservation

   d. are intended for property conservation and not necessarily life safety

134. You are preparing to hoist a pike pole to a second story roof. You are on the ground tying the rope. From your training, you remember that the two ways of tying a clove hitch are _____ or around an object.

   a. in the open

   b. in the air

   c. bowline

   d. Becket bend

135. In the absence of power tools, breeching a masonry wall begins _____.

   a. with a large blow delivered by a flat head axe

   b. with a large blow to the brick, followed by prying on the brick

   c. by removing mortar around the brick

   d. shaving the face of the brick to find the weak point and then applying a forceful blow

136. Your rookie firefighter is having trouble with terminology. You remind him that _____ are devices that water flows through.
    a. Higbees
    b. appliances
    c. hose tools
    d. hose bridges

137. Making equipment safe is addressed in three ways. Which of the following is not one of these three ways?
    a. Equipment selection
    b. Equipment application
    c. Equipment inspection and maintenance
    d. Equipment complexity

138. You are fighting a grass fire that is downhill from the position of the engine. You know that you will _____ head pressure in this operation.
    a. gain
    b. lose
    c. modify
    d. not need

139. You are responding to a wildland incident. While donning your boots, you break the shoelaces on one boot. What would your appropriate course of action be?
    a. Pull the lace out and tie it around the top of the boot to keep the boot on your leg.Request a lace from the quartermaster that evening.
    b. Borrow an extra lace or heavy string. Lace it through the eyelets and continue donning equipment. Replace both laces as soon as possible.
    c. Leave the boot loose, quickly don all PPE, and start hiking into the fire. Ask for shoelaces from coworkers on the way into the fire.
    d. Wrap heavy tape around your foot and ankle to keep the boot in place. Hike into the fire. Request a lace from the quartermaster that evening.

140. A _____ is the designation for a set number of resources of the same type and kind.
    a. crew
    b. strike team
    c. type
    d. section

141. You respond to a possible residential structure fire call and find that a teenager has burned popcorn in the microwave oven. There is no fire, but the smell of burnt popcorn smoke is throughout the first floor of the house. As a customer service, you should consider _____.
    a. taking the microwave to the dumpster
    b. PPV of the unaffected portions
    c. PPV of the affected portions
    d. removing all popcorn so the episode cannot be repeated

142. You have been assigned the duty of stopping the flow from an activated sprinkler head. You should _____.

    a. rely on past practice to stop the flow and establish a leak-free seal

    b. use a wedge or sprinkler tongs

    c. prepare to get seriously wet

    d. all of the above

143. Which form of transportation used STCC codes?

    a. Highway

    b. Rail

    c. Air

    d. Water

144. If a hose bursts during fire attack, and there is no clamp available, _____.

    a. a firefighter has no choice but to have the pump operator shut down the line

    b. the firefighter can fold the hose twice over itself and kneel down to hold pressure buildup in the kinks

    c. the firefighter can use an airpack to clamp off the line

    d. the hose can still be used, provided it is supplying sufficient psi at the nozzle

145. The acronym PASS stands for _____.

    a. Pull, Aim, Squeeze, Sweep

    b. Point, Aim, Squeeze, Sweep

    c. Point, Aim, Sight, Sweep

    d. Pull, Aim, Sight, Sweep

146. The maximum quantity of flammable and combustible liquids permitted to be stored inside a building _____.

    a. varies according to the occupancy

    b. is 1,000 gallons

    c. is 500 gallons

    d. is not a concern of firefighters conducting inspections

147. Your engine has parked at the scene of a structure fire. The driver operator tells you she is going to use a reverse lay. She is referring to _____.

    a. a direct attack

    b. an indirect attack

    c. a master stream

    d. water supply

148. You are using a fire extinguisher to fight a pile of wood that is on fire. You estimate there is a little more than one cubic foot of wood. Your extinguisher is a 2A10BC. You know you _____.

    a. will have to go for another extinguisher as the amount of agent in this extinguisher is inadequate

    b. have more than enough agent to combat this fire

    c. will have to go for another type of extinguisher to meet the ABC requirement

    d. will have to decon the area after the fire is out as you are using dry chemicals considered to be hazardous

149. You are called to a boat fire on a lake. Upon arrival, you find a large wooden boat burning.The boat is docked. There are no immediate fire hydrants visible. You should _____.

    a. use a reverse lay to the nearest hydrant

    b. request a water tender

    c. use booster tank water sparingly

    d. consider drafting

150. Many jurisdictions are working together to fight a wildfire on a mountain. The incident commander wishes to review the incident action plan. Who will she contact to get this information?

    a. The operations section chief

    b. The planning section chief

    c. The logistics section chief

    d. The finance/administration section chief

151. You respond to a reported smoke detector activation in the mainframe computer room of a school. Upon arrival, you are told there is a sprinkler system, but it has not activated. Based on this information, you can be relatively sure that the system is a _____ system.

    a. deluge

    b. dry pipe

    c. wet pipe

    d. pre-action

152. The use of the bowline knot has been greatly reduced because of _____.

    a. failure

    b. NFPA standardization of rope fibers

    c. synthetic rope

    d. natural fiber rope

153. In addition to the orange response guide section, a highlighted chemical in the yellow or blue section requires the reader to look in which section?

    a. Response guides

    b. Table of initial isolation and evacuations table

    c. Placard guides

    d. Truck and railcar guides

154. In the 1970s, the US government commissioned a panel of fire service experts to study the country's changing fire problem. As a result of the document this group published, many positive changes occurred in fire service at various levels of government. What was the name of the document that was produced?

    a. America Burning

    b. The National Commission on Fire Protection and Control Journal

    c. The United States Fire Administration Report

    d. Fire Protection and Control in the US

155. The _____, which is a part of the Department of Labor, is responsible for the enforcement of safety-related regulations in the workplace.

    a. National Fire Protection Association

    b. Code of Federal Regulations

    c. Occupational Safety and Health Administration

    d. United States Fire Administration

156. You are called to a boat fire on a lake. Upon arrival, you find a large wooden boat burning. The boat is docked. There are no immediate fire hydrants visible. You should _____.

    a. use a reverse lay to the nearest hydrant

    b. request a water tender

    c. use booster tank water sparingly

    d. consider drafting

157. If you are working a fireline on a wildland fire, and find yourself in the path of a fast-moving fire, you must ensure you have _____ before deploying your shelter.

    a. a shovel or pulaski available

    b. ear and eye protection

    c. no other options

    d. a radio

158. The potential hazards section in the DOT ERG response guides lists the _____.

    a. most dangerous hazard first

    b. most dangerous hazard last

    c. highlights the hazard

    d. is in random order

159. Which category of chemical WMD agents may present with delayed symptoms?

    a. Blister agents

    b. Nerve agents

    c. Explosives

    d. Irritants

160. You and your partner are assigned the task of advancing a hoseline to a second story window. Your partner takes an uncharged hoseline and begins ascent. For a safe operation, you should use _____ as a guide to begin your climb.

    a. the next coupling

    b. the middle of the next section

    c. the third loop

    d. the pre-marked section

161. The _____ section chief is responsible for documenting cost of materials and personnel for the incident.

    a. operations

    b. logistics

    c. planning

    d. finance/administration

162. You and your partner are advancing an attack line to a reported room and contents fire in an apartment. Upon reaching the door of the interior room where the fire is, you find that a couch at the far end of the room is burning freely, but there appears to be no real spread beyond the furniture. Using a(n) _____, you apply a fog pattern into the room and then close the door.

    a. aggressive straight stream attack

    b. direct attack

    c. combination attack

    d. indirect attack

163. You have been tasked with search and rescue on a residential structure fire. It is 3 A.M. The home is a two-story home. It can generally be assumed that your greatest rescue potential lies _____.

    a. on the uppermost floor

    b. in the basement

    c. on ground level

    d. both a and b

164. You are fighting a very large and hot room and contents fire in a residential structure. Because of poor visibility, your team decides to make a(n) _____. After this tactic, your team will probably have to back out of the fire area until the steam is dissipated.

    a. transitional attack

    b. indirect attack

    c. direct attack

    d. fire attack

165. What is combustion?

    a. A rapid, persistent chemical change that releases heat and light and is accompanied by flame

    b. Rapid oxidation with the development of heat and light

    c. A reaction that is a continuous combination of a fuel with certain elements, prominent among which is oxygen in either a free or combined form

    d. All of the above

166. A flammable vapor fire or explosion requires that _____.

    a. temperature of the liquid be very low

    b. proper air-to-fuel mixture

    c. it be above the UEL

    d. it be below the LEL

167. You are operating an engine with two hoselines in use. Your officer asks you to charge another line for exposure protection. You will know if this is or is not possible because you marked the _____ pressure before charging any lines.

    a. residual

    b. static

    c. flow

    d. none of the above

168. Mobile water supply apparatus, according to NFPA 1901, Standard for Automotive Fire Apparatus, must have a minimum of a _____ -gallon water tank.

    a. 500

    b. 750

    c. 1,000

    d. 1,500

169. You are recalled to the scene of a structure fire. You have been on vacation and have just run a marathon the previous day. When you don your SCBA, it is important to consider that _____.

    a. your legs may no longer support the additional weight of the SCBA

    b. you may have lost weight

    c. you may have forgotten important procedures on your vacation

    d. your partners have gotten used to fighting fire without you

170. The _____ detector measures temperature increases above a predetermined rate.

    a. progressive heat

    b. fixed temperature

    c. thermal increase

    d. rate-of-rise

171. You are holding a patient's head after extricating her from a motor vehicle accident. The patient complained of severe neck pain. Your partners lay a backboard out next to the patient.One rescuer assumes the position at the patient's torso, another at the patient's hips, and another at the patient's feet._____ will direct the movement of the patient onto the backboard.

    a. The rescuer at the torso

    b. You

    c. The rescuer at the feet

    d. The rescuer at the hips

172. The _____ nozzle was designed for aircraft operations.

    a. fog

    b. piercing

    c. ARFF ram

    d. straight-bore

173. You are on the scene of a multifamily residential structure fire in the middle of the night. A search team has found no occupants on the fire floor. Using the acronym REVAS, you should now begin _____ operation

    a. rescue

    b. evacuation

    c. ventilation

    d. exposure

174. Which of the following parts of a ladder can be found on an extension ladder but not on a straight ladder?

    a. A halyard

    b. Dogs, pawls, rung locks, or ladder locks

    c. A pulley

    d. All of the above

175. During a residential structure fire operation, there is a collapse of the interior first floor. You remind your partner that victims may survive _____.

    a. in the voids

    b. in the basement

    c. in the spans

    d. on the first floor

176. When you are working a fireline on a wildland fire, you wear leather boots that rise well above your ankle.This is for protection against _____.

    a. snake bites

    b. ankle injuries

    c. foot fatigue

    d. all of the above

177. What type of door can be used as a required exit in a building?

    a. Swinging door

    b. Revolving door

    c. Overhead door

    d. Sliding door

178. "Black fire" is apparent when a team of firefighters prepares to initiate fire attack. The smoke is high volume with a turbulent velocity and is ultradense and black. What should the firefighters expect to occur when they "read" this smoke condition?

    a. Backdraft

    b. Flash point

    c. Autoignition and flashover

    d. The decay phase

179. You are a venting a flat-roofed structure. After leaving the ladder, you should
_____.

    a. cut the vent hole

    b. notify command you are on the roof

    c. make a rectangular inspection cut

    d. make a triangular inspection cut

180. The STEL is an average of _____ for a chemical exposure.

    a. 15 minutes

    b. 8 hours

    c. 24 hours

    d. 1 hour

181. As hose is vital to firefighter operations, it is imperative that you remember that hose testing should occur _____.

    a. annually

    b. bi-annually

    c. daily

    d. visually during reloading

182. When you are using a BC fire extinguisher, you know _____.

    a. the agent and nozzle will not conduct electricity

    b. the agent will not conduct electricity, but the nozzle will

    c. the nozzle will not conduct electricity, but the agent will

    d. you are using the wrong extinguisher

183. You are checking your structural PPE and you discover a large rip in the vapor barrier of your coat. What should you do?

    a. Continue to use the PPE, but you should email your quartermaster that you will eventually need a replacement coat.

    b. Use vapor barrier adhesive to patch the liner and notify the department quartermaster.

    c. Replace the coat immediately.

    d. Ensure your duty uniform is worn beneath the jacket.

184. _____ fires are fires in combustibles metals, such as magnesium.

    a. Class A

    b. Class B

    c. Class C

    d. Class D

185. Which of the following is a role of a public safety telecommunicator?

    a. Receiving emergency requests from citizens

    b. Evaluating the need for public safety response

    c. Alerting responders to the scene of emergencies

    d. All of the above

186. _____ is/are a more predictable water source with less chance of contamination.

    a. Surface water

    b. Deep wells

    c. Natural springs

    d. Tanks, ponds, and cisterns

187. You have just been issued new work shirts from your quartermaster. You are unsure whether they have to be professionally cleaned or you can wash them yourself. The easiest way to answer your question is _____.

    a. to read the label

    b. to phone the quartermaster

    c. check the Internet

    d. read the proper NFPA standard

188. You have responded to a motor vehicle accident on a busy city street. Your SOPs state you should disconnect the battery of the car in all wrecks. You open the hood and find high voltage warnings all over the engine compartment. Nothing in the engine compartment looks familiar as many engine items are missing. You should assume the car is _____.

    a. radioactive

    b. hydraulic

    c. electric

    d. hybrid

189. You are the nozzle operator advancing a hoseline toward an automobile fire. The engine compartment is fully engulfed. You should approach the car from the _____.
    a. rear
    b. front
    c. side
    d. top

190. You and your partner have exited a residential structure fire after conducting a primary search. You return to your engine to change out your bottles and you notice your partner is quite pale. What should you do?
    a. Tell her what you see and encourage her to drink some water or Gatorade.
    b. Tell her what you see and have her sit in the shade for a few minutes while drinking fluid replacement drinks.
    c. Tell her what you see and suggest she report to rehab.
    d. Assume she knows her ability. If she does not complain, then it is safe to assume she is fine and ready for another assignment.

191. When hoisting small equipment, it is recommended to use a _____.
    a. water knot
    b. tag line
    c. bowline
    d. granny knot

192. Methane has a vapor density of 0.6. If Methane escapes its container, it can be expected to do what?
    a. Rise into the air
    b. Stay level with the container
    c. Collect in low-lying areas
    d. None of the above

193. Which of the following salvage tools would be helpful in a situation involving a flooded basement?
    a. Salvage covers
    b. Catch-alls
    c. Water vacuums
    d. Submersible pumps

194. You are attending church when a water motor gong sounds. As you are evacuating the building, you think and don't recall ever seeing an annunciator panel. Your reaction should be to _____.
    a. warn people to stay clear of the main exit as that is where firefighters will enter the structure
    b. phone in the alarm to dispatch once you have evacuated as you suspect it is a local alarm only
    c. grab a fire extinguisher and sweep the building looking for the fire
    d. gather all the children and hold them in one place for interrogation and lecture about setting off false alarms

195. The LEL for methane is 5%. When using a flammable gas detector calibrated for methane, the meter indicates a level of 100%. What is the amount of methane present by volume?
    a. 100%
    b. 50%
    c. 25%
    d. 5%

196. Firefighter Hernandez is getting ready to transmit a message on a mobile radio to the emergency communications center. Which of the following is false concerning the transmission of his message?
    a. The transmission must be within the time parameters established by the department or local jurisdiction.
    b. When he is ready, Firefighter Hernandez can hit the push-to-talk button and begin speaking immediately.
    c. Firefighter Hernandez should make sure the information he is transmitting is accurate, clear, and complete.
    d. Firefighter Hernandez should not attempt to transmit messages by radio while eating or chewing anything.

197. A backpack fire extinguisher and a pressurized water extinguisher _____.
    a. differ in that the backpack extinguisher is shaped to make wildland deployment easier
    b. differ in that the backpack extinguisher is cartridge operated
    c. differ in that the backpack extinguisher is hand operated
    d. have no differences

198. Hydraulics is the study of fluids _____.
    a. and their expansion capability
    b. and their effect on fire
    c. at rest and in motion
    d. none of the above

199. When transmitting messages on mobile and portable radios, firefighters should hold the microphone _____ inches from the mouth and at a _____ -degree angle for a clear transmission.
    a. 2-3; 45
    b. 1-2; 90
    c. 1-2; 45
    d. 2-3; 90

200. A Class A fire involves _____.
    a. ordinary combustibles
    b. flammable and combustible liquids, gases, and greases
    c. combustible metals and alloys
    d. energized electrical equipment

Final Exam: Answers to Questions

1. F	34. D	67. C	100. C
2. F	35. B	68. D	101. C
3. F	36. B	69. A	102. A
4. F	37. D	70. A	103. A
5. F	38. D	71. B	104. B
6. T	39. A	72. A	105. A
7. F	40. B	73. D	106. C
8. T	41. C	74. B	107. C
9. F	42. B	75. C	108. B
10. T	43. B	76. D	109. B
11. T	44. A	77. C	110. D
12. T	45. A	78. A	111. A
13. T	46. A	79. C	112. B
14. F	47. A	80. A	113. D
15. F	48. C	81. B	114. B
16. F	49. D	82. C	115. A
17. T	50. D	83. B	116. D
18. F	51. A	84. C	117. C
19. F	52. D	85. C	118. C
20. T	53. B	86. D	119. D
21. T	54. C	87. D	120. B
22. F	55. A	88. A	121. A
23. F	56. C	89. A	122. C
24. F	57. C	90. A	123. D
25. T	58. B	91. C	124. C
26. B	59. D	92. A	125. D
27. A	60. B	93. B	126. D
28. B	61. D	94. A	127. B
29. D	62. D	95. C	128. B
30. B	63. A	96. A	129. B
31. B	64. C	97. C	130. A
32. D	65. A	98. D	131. D
33. A	66. C	99. C	132. C

133. C	150. B	167. B	184. D
134. A	151. D	168. C	185. D
135. C	152. C	169. B	186. B
136. B	153. B	170. D	187. A
137. D	154. A	171. B	188. C
138. A	155. C	172. B	189. C
139. B	156. D	173. D	190. B
140. B	157. C	174. D	191. B
141. C	158. A	175. A	192. A
142. D	159. A	176. D	193. D
143. B	160. A	177. A	194. B
144. B	161. D	178. C	195. D
145. A	162. D	179. D	196. B
146. A	163. A	180. A	197. C
147. D	164. B	181. D	198. C
148. B	165. D	182. A	199. C
149. D	166. B	183. C	200. A

Final Exam: Rationale & References for Questions

Question #1. Here, depending on the time of day, the life hazard potential for fire victims is not as great, and the fire and life hazards to the firefighter can be greater. NFPA 1001: 5.3.10. *FFHB, 2E:* Page 615.

Question #2. In this situation, when applying the concept of risk/benefit, you can assume that the firefighters will not take significant risks to their safety to save what is already lost. This is especially true in light of the reports that there are no victims inside and very little property to save. NFPA 1001: 5.3; 6.3. *FFHB, 2E:* Page 730.

Question #3. Dressing a knot is the practice of making sure all parts of the knot are lying in the properorientation to the other parts and look exactly as the pictures indicated. NFPA 1001: 5.1.1.1. *FFHB, 2E:* Page 426.

Question #4. Figure 8-3 Older versions of fire extinguishers are labeled with colored geometrical shapes with letter designations. Ordinary combustibles are shown with a green triangle with the letter "A." NFPA 1001: 5.3.16. *FFHB, 2E:* Page 189.

Question #5. Detection systems are varied. Some require people to detect the fire and manually activate an alarm. Others are highly complex systems that can detect a fire almost at its ignition and sound an alarm. NFPA 1001: 6.5.1. *FFHB, 2E:* Page 309.

Question #6. When venting a series of windows, the firefighter must work toward the escape point. NFPA 1001: 5.3.11. *FFHB, 2E:* Page 581.

Question #7. Class I systems are designed for use by the fire department or trained personnel such as a fire brigade...No hose is provided for this class. NFPA 1001: 6.5.1. *FFHB, 2E:* Page 329.

Question #8. Every communications facility should be supported by a backup location in the event that the primary facility encounters problems that render it inoperable and result in evacuation. NFPA 1001: 5.2.1. *FFHB, 2E:* Page 50.

Question #9. Safety The water knot is the only knot that is recommended for use when tying webbing. NFPA 1001: 5.1.1.1. *FFHB, 2E:* Page 439.

Question #10. However, it is also important to practice the rescue knot on other people to simulate victims needing rescue. An emergency scene is not the proper venue for learning skills. NFPA 1001: 5.1.1.1. *FFHB, 2E:* Page 437.

Question #11. Damage from improper actions by firefighters include opening and closing valves or hydrants too quickly (thus creating a water hammer), failing to open or close a hydrant or other valve fully, or cross-threading the threads on the connections. NFPA 1001: 5.3.15. *FFHB, 2E:* Page 216.

Question #12. Emergency communications centers should be equipped with backup power supplies. NFPA 1001: 5.2.1. *FFHB, 2E:* Page 50.

Question #13. Firefighting is a team effort. Working alone or outside the action plan endangers individuals and the team. NFPA 1001: 6.1.1.1. *FFHB, 2E:* Page 119.

Question #14. It is more effective for firefighters to determine the type of construction and fire-resistive rating during a building's construction as opposed to viewing the building once it is complete. NFPA 1001: 5.3.10; 6.3.2. *FFHB, 2E:* Page 341.

Question #15. The call-taking process consistes of receiving a report, interviewing, and referral or dispatch composition. NFPA 1001: 5.2.1; 5.2.2. *FFHB, 2E:* Page 51.

Question #16. Hoselines can be hoisted whether charged or uncharged. NFPA 1001: 5.1.1.1. *FFHB, 2E:* Page 451.

Question #17. Because regular training is the single most important step in firefighter safety, the firefighter must strive to retain the information and skills that are presented in training sessions. NFPA 1001: 6.1.1.1. *FFHB, 2E:* Page 114.

Question #18. A firefighter has raised a ladder to a structure and is determining the proper distance the foot or butt of the ladder should be away from the building. The firefighter can estimate this by moving the foot of the ladder away from the building a distance of one-quarter the working length of the ladder. NFPA 1001: 5.3.6; 5.3.9; 5.3.10; 5.3.11; 5.3.12. *FFHB, 2E:* Page 400.

Question #19. Specific chemical information is important for a safe and informed response. NFPA 472: 5.4.4; 5.5.1. *FFHB, 2E:* Page 817.

Question #20. Advancing hoselines using a standpipe system involves two different hoseline evolutions. The first is the engine driver connecting to the fire department connection on the structure. The second is the hose crew connecting to the standpipe outlet and advancing the hoseline to attack the fire. NFPA 1001: 5.3. *FFHB, 2E:* Page 255.

Question #21. An understanding of the fire classes leads to selection of the proper unit and agent. NFPA 1001: 5.3.16. *FFHB, 2E:* Page 185.

Question #22. The two basic types of nozzles are solid stream (also called a smooth bore, straight bore, or solid tip) and fog nozzles with different styles available for each kind, especially fog nozzles. NFPA 1001: 5.3 NFPA 6.3. *FFHB, 2E:* Page 282.

Question #23. Fire hose needs to be tested prior to being placed in use and then retested annually during its lifetime. Hose also should be tested after being damaged and after repairs have been made. NFPA 1001: 6.5.3. *FFHB, 2E:* Page 274.

Question #24. A material that is heavy and dense will be a more effective thermal conductor than a material that is light and less dense. NFPA 1001: 5.3; 6.3. *FFHB, 2E:* Page 97.

Question #25. Incident readiness for firefighters includes personal protective equipment, personal accountability, and fit-for-duty status. NFPA 1001:5.3; 6.1.1.1; 6.3. *FFHB, 2E:* Page 723.

Question #26. Emergency evacuation plans should be drawn in graphic form and posted in each classroom and at various locations throughout the school. NFPA 1001: 5.5.1. *FFHB, 2E:* Page 682-683.

Question #27. The four elements of a fire stream are the pump, water, hose, and nozzle. NFPA 1001: 5.3 NFPA 6.3. *FFHB, 2E:* Page 281.

Question #28. Static pressure is the pressure in the system with no hydrants or water flowing. NFPA 1001: 5.3.15. *FFHB, 2E:* Page 214.

Question #29. First responders should have a basic awareness of the threat of terrorism and basic response actions. NFPA 472: 4.2.1. *FFHB, 2E:* Page 929.

Question #30. The webbing drag enables a rescuer who is significantly smaller than the victim to perform a rescue. NFPA 1001: 5.3.9. *FFHB, 2E:* Page 472.

Question #31. Ladders should be stored with multiple support points to prevent bowing or sagging. NFPA 1001: 5.3.6; 5.3.9; 5.3.10; 5.3.11; 5.5.3. *FFHB, 2E:* Page 381.

Question #32. A reverse lay is the opposite of the forward lay with the supply line being dropped off at the fire location and the engine laying the hose toward the water source. NFPA 1001: 5.3.10. *FFHB, 2E:* Page 270.

Question #33. Attack hose is used by trained personnel to fight fires. NFPA 1001: 5.3.15. *FFHB, 2E:* Page 222.

Question #34. From the seated position, it is extremely difficult, if not impossible, to check the cylinder gauge and compare it to the regulator gauge. NFPA 1001: 5.3.1. *FFHB, 2E:* Page 163.

Question #35. Understanding toxicology and health effects is important. NFPA 472: 5.2.2; 5.2.3. *FFHB, 2E:* Page 838.

Question #36. Fire hose is made in three types of construction: wrapped, braided, and woven. NFPA 1001: 5.3. *FFHB, 2E:* Page 221.

Question #37. To disconnect the opener, break out a panel near the attachment mechanism, reach in with a tool to grab the release cord, and pull. NFPA 1001: 5.3.4; 6.3.2. *FFHB, 2E:* Page 525.

Question #38. Dynamic influences such as dirty gear, moisture (including perspiration), and fabric compression (from an SCBA) can reduce the TPP of the clothing. NFPA 1001: 5.1.1.2. *FFHB, 2E:* Page 128.

Question #39. She should use the command "raise" as this means to move a ladder from a horizontal position to a vertical angle. NFPA 1001: 5.3.6; 5.3.9; 5.3.10; 5.3.11; 5.3.12. *FFHB, 2E:* Page 375.

Question #40. As an incident expands, branches are established to maintain span of control over a number of divisions, sectors, and/or groups. NFPA 1001: 6.1.1.1. *FFHB, 2E:* Page 42.

Question #41. Spanner wrenches come in several sizes and are used to tighten or loosen couplings. NFPA 1001: 5.1.1.1. *FFHB, 2E:* Page 227.

Question #42. Firefighter Jong would need to look in the Code of Federal Regulations, which is a set of documents that includes all federally promulgated regulations for all federal agencies. NFPA 1001: 6.1.1.1. *FFHB, 2E:* Page 109.

Question #43. An abrasion is a scrape or brush of the skin usually making it reddish in color and resulting in minor capillary bleeding. NFPA 1001: 4.3. *FFHB, 2E:* Page 712.

Question #44. A standard wildland hose load is the modified Gasner bar pack, which provides ease of rolling and stretching the line, convenience of carrying with hands free, and protection for the couplings. NFPA 1001: 5.1.1.1. *FFHB, 2E:* Page 250.

Question #45. From there, the team will work up one of the flanks of the fire working toward the head or progressive end of the fire. NFPA 1001: 5.3.19. *FFHB, 2E:* Page 621.

Question #46. Do not coil the rope in the bag; it will hang-up almost every time if you do. NFPA 1001: 5.1.1.1. *FFHB, 2E:* Page 447.

Question #47. Nozzle reaction is the force of nature that makes the nozzle move in the opposite direction of the water flow. NFPA 1001: 5.3 NFPA 6.3. *FFHB, 2E:* Page 282.

Question #48. Within an Incident Management System, there are four section chiefs: operations, logistics, planning, and finance/administration. NFPA 1001: 6.1.1.1. *FFHB, 2E:* Page 39-40.

Question #49. Extinguishers in buildings should be checked every 30 days, and extinguishers on apparatus should be inspected each time the vehicle is inspected. NFPA 1001: 5.3.16. *FFHB, 2E:* Page 198.

Question #50. Fire station tours should be tailored to fit the needs and interests of the visitors. The preschool children would be most receptive to a discussion on how to call for the fire department, as discussions about response procedures, detector placement, and first aid are beyond their comprehension level. NFPA 1001: 5.5.2. *FFHB, 2E:* Page 684.

Question #51. A rehabilitation area at the fireground is a reactive effort (intervention) to prevent injury. NFPA 1001: 6.1.1.1. *FFHB, 2E:* Page 110-111.

Question #52. Dump sites should be selected for availability to unload multiple tenders, turnaround area for the tenders, operational area, continued access to the fireground, and safety of personnel. NFPA 1001: 5.3.15. *FFHB, 2E:* Page 211.

Question #53. Installed fire-extinguishing systems such as total flooding carbon dioxide or halon systems create an oxygen-deficient atmosphere. NFPA 1001: 5.3.1. *FFHB, 2E:* Page 145.

Question #54. Ventilate as you advance, as long as it will not spread the fire. Ventilation allows the products of combustion to escape and provides a better environment. Check for outside openings such as windows and doors. This will provide a means of escape in an emergency and provide the firefighters' location to outside personnel. NFPA 1001: 5.3.1. *FFHB, 2E:* Page 150.

Question #55. Extinguishers in buildings should be checked every 30 days, and extinguishers on apparatus should be inspected each time the vehicle is inspected. NFPA 1001: 5.3.16. *FFHB, 2E:* Page 198.

Question #56. A leg lock is used to secure the firefighter to the ladder when both hands must be used to perform a task and a ladder belt is not available. NFPA 1001: 5.36; 5.3.9; 5.3.10; 5.3.11; 5.3.12. *FFHB, 2E:* Page 408.

Question #57. Vapor pressure is the measurable amount of pressure being exerted against a confined container by a liquid substance as it converts to a gas. NFPA 1001: 5.3; 6.3. *FFHB, 2E:* Page 86.

Question #58. The positive-pressure technique actually injects air into the compartment and pressurizes it. NFPA 1001: 5.3.11. *FFHB, 2E:* Page 573.

Question #59. Firefighters should don all PPE necessary for the potential worst-case scenario. Granted, this approach may lead to "overdressing" for an incident. In these cases, the firefighters' company officer, incident commander, or incident safety officer may allow firefighters to "dress-down." NFPA 1001: 5.1.1.2. *FFHB, 2E:* Page 137.

Question #60. A water vacuum is a commonly used tool in salvage operations. NFPA 1001: 5.3.14. *FFHB, 2E:* Page 635.

Question #61. Understanding various protective actions is important for responder health and safety. NFPA 472: 5.4.2. *FFHB, 2E:* Page 859.

Question #62. Reports of emergencies can be received in a variety of ways, including conventional telephones, wireless or cellular telephones, emergency call boxes, automatic alarms, TDD equipment, still alarms, or walk-ups NFPA 1001: 5.2.1; 5.2.2. *FFHB, 2E:* Page 53-54.

Question #63. Firefighters are responding to a fire in a structure that is Type V construction. The most common occupancy of this type of construction is residential. NFPA 1001: 5.3.10; 6.3.2. *FFHB, 2E:* Page 360.

Question #64. The logistics section chief is responsible for securing the facilities, services, equipment, and materials for an incident. NFPA 1001: 6.1.1.1. *FFHB, 2E:* Page 40.

Question #65. The maximum load for a ladder can be found on the manufacturer's label affixed to the ladder. NFPA 1001: 5.3.6; 5.3.9; 5.3.10; 5.3.11; 5.3.12. *FFHB, 2E:* Page 392.

Question #66. The accident chain includes the following components: the environment, human factors, equipment, the event, and the injury. NFPA 1001: 6.1.1.1. *FFHB, 2E:* Page 110.

Question #67. Attach locking-type pliers to the lock case and lock the jaws. The locking pliers must have a chain or rope attached so the firefighter can hold the pliers clear of the saw. A rotary saw with metal cutting blade is used to cut the lock shackles. Figure 17-50 Cut both sides of the shackle on one cut. NFPA 1001: 5.3.4; 6.3.2. *FFHB, 2E:* Pages 540-541.

Question #68. Common sense dictates that firefighters should use SCBA on every fire scene—from start to finish. NFPA 1001: 5.3.1. *FFHB, 2E:* Page 148.

Question #69. Understanding toxicology and health effects is important. NFPA 472: 5.2.2; 5.2.3. *FFHB, 2E:* Page 838.

Question #70. The bubbles are filled with a gas, usually air, creating a blanket over the surface of the fuel to cool and smother the fire, while sealing the escaping vapors. NFPA 1001: 5.3 NFPA 6.3. *FFHB, 2E:* Page 294.

Question #71. Check for proper seal by attaching the regulator and inhaling a breath, activating the regulator.Hold that breath for about five seconds and listen and feel for any air leaks. NFPA 1001: 5.3.1. *FFHB, 2E:* Page 166.

Question #72. NFPA 1581, Standard on Fire Department Infection Control Program, requires that clothing be cleaned every six months as a minimum. NFPA 1001: 5.1.1.2. *FFHB, 2E:* Page 136.

Question #73. Pre-action systems are used in areas where the materials protected are of high value and water damage would be expensive, such as computer rooms and historical items. NFPA 1001: 6.5.1. *FFHB, 2E:* Page 321.

Question #74. Occasionally water should be flowed through stored sections to prevent the liner from drying and cracking, and then it should be re-dried and stored. NFPA 1001: 5.5.4. *FFHB, 2E:* Page 224.

Question #75. Piled storage is held up by the integrity of the items stored, and as that changes with either fire or water damage, those piles may topple. NFPA 1001: 5.3.10. *FFHB, 2E:* Page 627.

Question #76. Daily maintenance: SCBA units should be checked daily to ensure they are secured and ready for operation. Monthly Maintenance: The monthly SCBA check contains all elements of the daily check, but adds several checks of the mechanics of the system. Annual and Bi-annual Maintenance: NIOSH and SCBA manufacturers require a number of different functional tests of SCBA units. NFPA 1001: 5.3.1. *FFHB, 2E:* Page 172.

Question #77. These valves must have a chain lock on them to prevent tampering. A wrench or a wheel controls these valves; a padlock and chain are used to lock them open. NFPA 1001: 6.5.1. *FFHB, 2E:* Pages 324 and 325.

Question #78. Firefighters must never enter a total flooding environment without first engaging their SCBA. NFPA 1001: 6.5.1. *FFHB, 2E:* Page 336.

Question #79. Company officers are supervisory-level positions and are responsible for both firefighters and administrative duties. NFPA 1001: 5.1.1.1. *FFHB, 2E:* Page 28.

Question #80. Risk management is the process of minimizing the chance, degree, or probability of damage, loss, or injury. NFPA 1001: 6.1.1.1. *FFHB, 2E:* Page 108.

Question #81. This system allows the pump to operate at full flow, but may cause damage to the pump and valves without proper back-flushing of the entire pump. NFPA 1001: 5.3 NFPA 6.3. *FFHB, 2E:* Page 298.

Question #82. Spanner wrenches come in several sizes and are used to tighten or loosen couplings. NFPA 1001: 5.1.1.1. *FFHB, 2E:* Page 227.

Question #83. Standpipe systems are designed to allow firefighters to fight fires in larger buildings by pre-piping water lines for fire streams throughout the building. NFPA 1001: 6.5.1. *FFHB, 2E:* Page 328.

Question #84. During fire safety inspections, firefighters should check fire sprinkler systems to ensure that all water supply vales are open and secured. NFPA 1001: 5.5.1. *FFHB, 2E:* Page 666.

Question #85. Based on the scene assessment, the first-arriving apparatus should be positioned to create a traffic barrier to help shield the greatest number of rescuers. NFPA 1001: 6.4.1. *FFHB, 2E:* Page 487.

Question #86. Dry chemical or dry powder extinguishers carried on apparatus should occasionally be rotated upside down and shaken to keep powders from packing at the bottom as a result of vehicle vibration. NFPA 1001: 5.3.16. *FFHB, 2E:* Page 197.

Question #87. EP=NP+FL±E+SA NFPA 1001: 5.3 NFPA 6.3. *FFHB, 2E:* Page 92.

Question #88. The Becket bend is utilized to tie ropes of equal diameter together, while the double Becket bend is used most often when tying ropes of unequal diameter. NFPA 1001: 5.1.1.1. *FFHB, 2E:* Page 429.

Question #89. Static pressure is the pressure in the system with no hydrants or water flowing. The pump operator then charges the first line with the desired volume, noting the pressure first. With the flow going, the operator again reads the intake gauge and gets the residual pressure or the remaining pressure left in the system after the flow and friction loss from the flow. The pump operator then compares the percentage of pressure drop from static to residual and determines the amount of additional volumes that may be pumped from that hydrant. NFPA 1001: 5.3.15. *FFHB, 2E:* Page 214.

Question #90. The Becket bend is utilized to tie ropes of equal diameter together, while the double Becket bend is used most often when tying ropes of unequal diameter. NFPA 1001: 5.1.1.1. *FFHB, 2E:* Page 429.

Question #91. "Black fire" is apparent when a team of firefighters prepares to initiate fire attack. The smoke is high volume with a turbulent velocity and is ultradense and black. They should expect autoignition or flashover when they "read" this smoke condition. NFPA 1001: 5.3; 6.3. *FFHB, 2E:* Page 103.

Question #92. The follow-through figure-eight knot is very useful when attaching a utility or life-safety line rope to an object that does not have a free end available. NFPA 1001: 5.1.1.1. *FFHB, 2E:* Page 433.

Question #93. For instance, a 2-A extinguisher will put out twice the fire of a 1-A. NFPA 1001: 5.3.16. *FFHB, 2E:* Page 194.

Question #94. An oxidizer is a catalyst in the breakdown of molecules and possesses a chemical property that can pull apart a molecule and break apart the bond that previously existed. NFPA 1001: 5.3; 6.3. *FFHB, 2E:* Page 81.

Question #95. Flashover occurs when the entire contents of a compartment ignite almost simultaneously, generating intense heat and flames. NFPA 1001: 5.3; 6.3. *FFHB, 2E:* Page 99.

Question #96. In addition to the aluminized fabric, proximity PPE features full face shields that are coated with an anodized gold material to help create a "mirrored" reflective surface. Without this special coating, the wearer could receive radiant facial burns, and the face shield could quite possibly melt. NFPA 1001: 5.1.1.2. *FFHB, 2E:* Page 130.

Question #97. The rehabilitation area is where firefighters can rest and rehydrate after performing fireground activities. NFPA 1001:5.3; 6.3. *FFHB, 2E:* Page 730-731.

Question #98. EP=NP+FL±E+SA NFPA 1001: 5.3 NFPA 6.3. *FFHB, 2E:* Page 92.

Question #99. When connecting to a Siamese, the outlet on the far left should be chosen first because this will allow better access for using the spanner wrench to tighten the coupling. Some connections only have clappers installed on the right side. NFPA 1001: 6.5.1. *FFHB, 2E:* Page 324.

Question #100. Residential sprinklers are designed for life safety and not necessarily to protect property. NFPA 1001: 6.5.1. *FFHB, 2E:* Page 314.

Question #101. Any irregularities should be noted and repaired, or the SCBA should be pulled from service until a department technician can repair the unit. NFPA 1001: 5.3.2. *FFHB, 2E:* Page 172.

Question #102. Transfer of command should occur during a face-to-face meeting; but under extreme conditions, transfer may be accomplished by radio or telephone. NFPA 1001: 6.1.1.1. *FFHB, 2E:* Page 37.

Question #103. Understanding protective clothing and its relationship to toxicology and health effects is important. NFPA 472: 5.3.3; 5.4.3. *FFHB, 2E:* Page 843.

Question #104. If the fire is already in the space, the tiles can usually be unseated with the hoseline merely by directing the stream upward and extinguishing fire in the ceiling while advancing. NFPA 1001: 5.3.12. *FFHB, 2E:* Page 587.

Question #105. Specific chemical information is important for a safe and informed response. NFPA 472: 4.2.1 - 4.2.2. *FFHB, 2E:* Page 826.

Question #106. Hydraulic pistons are used on aerial ladder devices to raise and lower the ladder. They are not found on ground ladders. NFPA 1001: 5.3.6; 5.3.9; 5.3.10; 5.3.11; 5.5.3. *FFHB, 2E:* Pages 372-374.

Question #107. A firefighter discovers that a facility with outdoor storage of hazardous materials has a six-inch curb around the storage area. This is called secondary containment and is meant to contain the stored liquid if it should escape from its container. NFPA 1001: 5.5.1. *FFHB, 2E:* Page 670.

Question #108. The firefighter should be positioned to prevent being pinned between the hose and the hydrant. At this point, the engine can move forward completing the hose lay. NFPA 1001: 5.3. *FFHB, 2E:* Page 262.

Question #109. The public information officer is a command staff position and reports directly to the incident commander. NFPA 1001: 6.1.1.1. *FFHB, 2E:* Page 41.

Question #110. Now firefighters can choose from traditional rubber-like boots or leather-type boots. Each has advantages and disadvantages. NFPA 1001: 5.1.1.2. *FFHB, 2E:* Page 129.

Question #111. First responders should have a basic awareness of the threat of terrorism and basic response actions. NFPA 472: 4.2.1. *FFHB, 2E:* Page 917.

Question #112. A swimming pool at a residential occupancy indicates storage of chemicals such as chlorine, which will produce a poisonous gas. Firefighters must understand that there is no "routine" fire. Even a light smoke condition at this type of structure could be deadly. NFPA 1001: 5.3.1. *FFHB, 2E:* Page 146.

Question #113. All these methods can be used to control external bleeding. NFPA 1001: 4.3. *FFHB, 2E:* Page 711-712.

Question #114. Career firefighters banded together near the beginning of the twentieth century to form a labor union, the International Association of Firefighters. This was not during the Civil War period. NFPA 1001: N/A. *FFHB, 2E:* Page 13-15.

Question #115. PPE provides a minimum level of protection and should be considered the last resort of protection for firefighters and emergency responders operating at an incident. NFPA 1001: 5.1.1.2. *FFHB, 2E:* Page 125.

Question #116. Personal size-up can be defined as a continuous mental evaluation of firefighters' immediate environments, facts, and probabilities. NFPA 1001:5.3; 6.3. *FFHB, 2E:* Page 725.

Question #117. Is a fog stream the correct choice?If so, which width? Figure 19-19 Firefighters use a fog pattern to shield themselves from radiant heat. NFPA 1001: 5.3.10. *FFHB, 2E:* Page 604.

Question #118. Firefighters arrive on a scene to find a Class C fire. The best method of attack is to remove the flow of electricity and then re-analyze the class of the fire for further actions. A Class A or B fire may be left after the Class C fire is resolved. NFPA 1001: 5.3; 6.3. *FFHB, 2E:* Page 100-101.

Question #119. The running end is used for work such as hoisting a tool. NFPA 1001: 5.1.1. *FFHB, 2E:* Page 425.

Question #120. A distinct advantage of the PPV blower is ease of setup. If properly applied, the blower can do the job of several firefighters attempting to open multiple holes with ladders at different locations. NFPA 1001: 5.3.11. *FFHB, 2E:* Page 565.

Question #121. For example, opening a bulkhead door will be quick and effective. It should be performed before a roof cut is initiated. NFPA 1001: 5.3.11. *FFHB, 2E:* Page 576.

Question #122. When AFFF was introduced, one of its additional advantages was that it did not require a special foam nozzle for application. NFPA 1001: 5.3 NFPA 6.3. *FFHB, 2E:* Page 302.

Question #123. The information provided to an emergency communications center by mobile or portable radio must be accurate, clear, and complete. NFPA 1001: 5.2.3. *FFHB, 2E:* Page 67.

Question #124. The shoulder toss is done by one firefighter and is used for covering large, unbreakable items. NFPA 1001: 5.3.14. *FFHB, 2E:* Page 643-644.

Question #125. As previously mentioned, the laid method is the most common type of construction for natural fiber ropes. NFPA 1001: 5.1.1.1. *FFHB, 2E:* Page 422.

Question #126. A communications center may also be referred to as a PSAP, or public safety answering point. NFPA 1001: 5.2.1. *FFHB, 2E:* Page 48.

Question #127. The simple act of practicing hands-on tool use with gloves on can increase muscle memory and reduce the frustration of lost dexterity. NFPA 1001: 5.1.1.2. *FFHB, 2E:* Page 129.

Question #128. The original Halligan tool, designed by Hugh Halligan of the Fire Department of the City of New York, has proven to be the most important single forcible entry tool used in the fire service. NFPA 1001: 5.3.4. *FFHB, 2E:* Page 513.

Question #129. The biggest problem with PASS devices results when wearers simply forget to turn their units on. This simple mental lapse has contributed to numerous firefighter fatalities. NFPA 1001: 5.1.1.2. *FFHB, 2E:* Page 135.

Question #130. The wildland PPE ensemble is designed to be worn over undergarments. These undergarments (long-sleeve t-shirt, pants, and socks) should be 100 percent cotton or of a fire-resistive material. NFPA 1001: 5.1.1.2. *FFHB, 2E:* Page 130.

Question #131. In addition to voice communications, enhanced 9-1-1 service provides emergency communications centers with the telephone number and address of the phone from which the call is originating. NFPA 1001: 5.2.1; 5.2.2. *FFHB, 2E:* Page 54.

Question #132. Master streams or heavy appliances are non-hand-held water applicators capable of flowing more than 350 gallons of water per minute. The wagon pipe is a permanently mounted master stream device on an engine that has either pre-piped water connection or needs a short section of hose to connect it to the pump. Some departments refer to wagon pipes as a "deck gun." NFPA 1001: 5.3. *FFHB, 2E:* Page 271.

Question #133. Residential sprinklers are designed for life safety and not necessarily to protect property. NFPA 1001: 6.5.1. *FFHB, 2E:* Page 314.

Question #134. NFPA 1001: 5.1.1.1. *FFHB, 2E:* Page 238.

Question #135. Work at the mortar joints because this is usually the weak point. NFPA 1001: 5.3.4; 6.3.2. *FFHB, 2E:* Page 544.

Question #136. Appliances are devices that water flows through, including adapters and connectors. NFPA 1001: 5.3.15. *FFHB, 2E:* Page 227.

Question #137. Making equipment safe is addressed in three ways: equipment selection, equipment inspection and maintenance, and equipment application. NFPA 1001: 5.5.3; 5.5.4; 6.1.1.1; 6.5.2. *FFHB, 2E:* Page 112.

Question #138. Head pressure measures the pressure at the bottom of a column of water in feet. Head pressure can be gained or lost when water is being pumped above or below the level of the pump. NFPA 1001: 5.3 NFPA 6.3. *FFHB, 2E:* Page 291.

Question #139. Additionally, a good fitting, tightly laced boot can help prevent ankle sprains and reduce foot fatigue. NFPA 1001: 5.1.1.2. *FFHB, 2E:* Page 132.

Question #140. A strike team is the designation for a set number of resources of the same type and kind. NFPA 1001: 6.1.1.1. *FFHB, 2E:* Page 42.

Question #141. The positive-pressure technique actually injects air into the compartment and pressurizes it. In an attempt to equalize, the smoke and heat are carried out into the areas of lower pressure outside the structure. NFPA 1001: 5.3.11. *FFHB, 2E:* Page 573.

Question #142. While it may seem easy, stopping water flow from a sprinkler head requires practice to effectively stop the flow and establish a leak-free seal. The firefighter assigned to stop sprinkler flow at the head will get seriously wet. NFPA 1001: 6.5.1. *FFHB, 2E:* Page 326.

Question #143. Specific chemical information is important for a safe and informed response. NFPA 472: 4.2.1 - 4.2.2. *FFHB, 2E:* Page 825.

Question #144. If no clamp is available, a firefighter can fold the hose twice over itself and kneel down to hold pressure buildup in the kinks. NFPA 1001: 5.3.10. *FFHB, 2E:* Page 269.

Question #145. PASS, the four steps for using a fire extinguisher:Pull the pin. Aim the nozzle. Squeeze the handle. Sweep the base of the fire. NFPA 1001: 5.3.16. *FFHB, 2E:* Page 195.

Question #146. The maximum quantity of flammable and combustible liquids permitted to be stored inside a building varies according to the occupancy. NFPA 1001: 5.5.1. *FFHB, 2E:* Page 669.

Question #147. A reverse lay is the opposite of the forward lay with the supply line being dropped off at the fire location and the engine laying the hose toward the water source. NFPA 1001: 5.3.10. *FFHB, 2E:* Page 270.

Question #148. For a 1-A rating, the extinguisher should extinguish a wood crib fire of about one cubic foot.The ratings increase as the amount of fire suppressed increases. For instance, a 2-A extinguisher will put out twice the fire of a 1-A. NFPA 1001: 5.3.16. *FFHB, 2E:* Page 194.

Question #149. Other natural sources of surface water are rivers, lakes, and ponds. Drafting The pumping of water from a static source by taking advantage of atmospheric pressure to force water from the source into the pump NFPA 1001: 5.3.15. *FFHB, 2E:* Page 204.

Question #150. The planning section chief is responsible for the development of the incident action plan. NFPA 1001: 6.1.1.1. *FFHB, 2E:* Page 39.

Question #151. Pre-action systems are used in areas where the materials protected are of high value and water damage would be expensive, such as computer rooms and historical items. NFPA 1001: 6.5.1. *FFHB, 2E:* Page 321.

Question #152. Although it was the mainstay of the fire service knots for years, the advent of synthetic fiber ropes has greatly reduced the utilization of the bowline knot. NFPA 1001: 5.1.1.1. *FFHB, 2E:* Page 431.

Question #153. Specific chemical information is important for a safe and informed response. NFPA 472: 4.2.1 - 4.2.2. *FFHB, 2E:* Page 812.

Question #154. In the 1970's, the US government commissioned a panel of fire service experts to study the country's changing fire problem. This group was called the National Commission on Fire Protection and Control. As a result of the document this group published, America Burning, many positive changes occurred in fire service at various levels of government, including, at the federal level, the creation of the United State Fire Administration and the National Fire Academy. NFPA 1001: N/A. *FFHB, 2E:* Page 18.

Question #155. The Occupational Safety and Health Administration, which is a part of the Department of Labor, is responsible for the enforcement of safety-related regulations in the workplace. NFPA 1001: 6.1.1.1. *FFHB, 2E:* Page 109.

Question #156. Other natural sources of surface water are rivers, lakes, and ponds. Drafting The pumping of water from a static source by taking advantage of atmospheric pressure to force water from the source into the pump NFPA 1001: 5.3.15. *FFHB, 2E:* Page 204.

Question #157. A fire shelter is a last-resort protective device for firefighters caught or trapped in an environment where a firestorm or blowup is imminent. NFPA 1001: 1977. *FFHB, 2E:* Page 132.

Question #158. Specific chemical information is important for a safe and informed response. NFPA 472: 4.2.1 - 4.2.2. *FFHB, 2E:* Page 812.

Question #159. First responders should have a basic awareness of the threat of terrorism and basic response actions. NFPA 472: 4.2.1. *FFHB, 2E:* Page 929.

Question #160. The next firefighter places the hoseline over the left shoulder at the next coupling and begins to climb the ladder. NFPA 1001: 5.3. *FFHB, 2E:* Page 259.

Question #161. The finance/administration section chief is responsible for documenting cost of materials and personnel for the incident. NFPA 1001: 6.1.1.1. *FFHB, 2E:* Page 40.

Question #162. Indirect fire attack is used to attack interior fires by applying a fog stream into a closed room or compartment, converting the water into steam to extinguish the fire. Firefighters apply the water at the doorway and then close the door, allowing the steam to put the fire out. NFPA 1001: 5.3 NFPA 6.3. *FFHB, 2E:* Page 289.

Question #163. In the two-story home, the bedrooms will generally be on the upper floor (not always) with possibly one of them downstairs. NFPA 1001: 5.1.1.1. *FFHB, 2E:* Page 612.

Question #164. This method, if used indoors, will usually greatly disturb the thermal balance, cause a loss of visibility in the structure as heat and products of combustion are circulated downward. When using this method, the firefighting team should attempt to shield themselves from the steam vapor by backing out of the area after applying the water. They can reenter as soon as the vapor begins to dissipate. NFPA 1001: 5.3.10. *FFHB, 2E:* Page 604.

Question #165. All of these definitions have been used to describe combustion. NFPA 1001: 5.3; 6.3. *FFHB, 2E:* Page 76.

Question #166. Understanding of basic chemical and physical properties is important for the health and safety of emergency responders. NFPA 472: 5.2.2 - 5.2.4. *FFHB, 2E:* Page 800.

Question #167. Prior to charging any lines, the static pressure is read on the main intake compound gauge.The pump operator then charges the first line with the desired volume, noting the pressure first.With this flow going, the operator again reads the intake gauge and gets the residual pressure or the remaining pressure left in the system after the flow and friction loss from the flow. NFPA 1001: 5.3.15. *FFHB, 2E:* Page 215.

Question #168. Mobile water supply apparatus, according to NFPA 1901, Standard for Automotive Fire Apparatus, must have a minimum of a 1000-gallon water tank. NFPA 1001: 5.3.15. *FFHB, 2E:* Page 31.

Question #169. Even though face pieces have been fitted and tested, weight loss or a 24-hour growth of facial hair may affect the ability to obtain a good seal. NFPA 1001: 5.3.1. *FFHB, 2E:* Page 150.

Question #170. The rate-of-rise heat detector measures temperature increases above a predetermined rate. NFPA 1001: 6.5.1. *FFHB, 2E:* Page 310.

Question #171. When ready, the rescuer at the head directs the others to "roll patient." NFPA 1001: 4.3. *FFHB, 2E:* Page 479.

Question #172. Piercing nozzles were originally designed to penetrate the skin of aircraft and now have been modified to pierce through buildings' walls and floors. NFPA 1001: 5.3 NFPA 6.3. *FFHB, 2E:* Page 286.

Question #173. Still another acronym is termed REVAS: Rescue, Exposures, Ventilation, Attack, and Salvage. NFPA 1001: 5.3.3. *FFHB, 2E:* Page 607.

Question #174. A halyard, a pulley, and dogs (pawls, rung locks, ladder locks) are found on an extension ladder but not on a straight ladder. NFPA 1001: 5.3.6; 5.3.9; 5.3.10; 5.3.11; 5.5.3. *FFHB, 2E:* Pages 372-373.

Question #175. Voids Spaces within a collapsed area that are open and may be an area where someone could survive a building collapse. NFPA 1001: 6.4.2. *FFHB, 2E:* Page 506.

Question #176. Lace-up leather boots that rise well above the ankle (8 to 10 inches) help protect the wearer from cuts, snakebites, and burns. Additionally, a good fitting, tightly laced boot can help prevent ankle sprains and reduce foot fatigue. NFPA 1001: 5.1.1.2. *FFHB, 2E:* Page 132.

Question #177. Only a swinging door can be used as a required exit in a building. NFPA 1001: 5.5.1. *FFHB, 2E:* Page 664.

Question #178. "Black fire" is apparent when a team of firefighters prepares to initiate fire attack. The smoke is high volume with a turbulent velocity and is ultradense and black. They should expect autoignition or flashover when they "read" this smoke condition. NFPA 1001: 5.3; 6.3. *FFHB, 2E:* Page 103.

Question #179. The first operation that needs to be accomplished on a flat roof is the inspection cut. Then the triangular inspection cut is completed with another cut. NFPA 1001: 5.3.11. *FFHB, 2E:* Page 580.

Question #180. Understanding toxicology and health effects is important. NFPA 472: 5.2.2; 5.2.3. *FFHB, 2E:* Page 838.

Question #181. The testing of hose begins with a visual inspection of the hose coupling. This inspection should be done with the annual test and during routine reloading and reconnection of hose sections. As the hose is loaded, it should also be visually inspected for any type of damage. NFPA 1001: 6.5.3. *FFHB, 2E:* Page 274.

Question #182. Class C extinguishers have extinguishing agents and hoses with nozzles that will not conduct electricity. NFPA 1001: 5.3.16. *FFHB, 2E:* Page 186.

Question #183. Both components rely on a layered protection system that includes a fire-resistive outer shell, vapor barrier, and thermal barrier. NFPA 1001: 5.1.1.2. *FFHB, 2E:* Page 128.

Question #184. Class D fires are fires in combustibles metals, such as magnesium. NFPA 1001: 5.3; 6.3. *FFHB, 2E:* Page 100.

Question #185. The role of the public safety telecommunicator is to receive emergency requests from citizens, evaluate the need for public safety response, and alert responders to the scene of emergencies. NFPA 1001: 5.2.1. *FFHB, 2E:* Page 48.

Question #186. Deep wells may penetrate through several layers of water before finding an aquifer. They are a more predictable water source with less chance of contamination. NFPA 1001: 5.3.15. *FFHB, 2E:* Page 204.

Question #187. The key to maintaining personal protective equipment in a high state of readiness is simple: Follow the specific instructions given by the manufacturer. NFPA 1001: 5.1.1.2. *FFHB, 2E:* Page 136.

Question #188. Instead, the firefighter will find high voltage warning labels on components. NFPA 1001: 5.3.7. *FFHB, 2E:* Page 600.

Question #189. Firefighters should understand that when heated, as in an automobile fire, these hydraulic fluid-filled bumper systems undergo great stress as the fluid expands. This stress can cause the bumper to be propelled off the car, traveling up to 40 feet or more. Persons standing in front of the bumper (front or rear) when this happens can be severely injured. NFPA 1001: 5.3.7. *FFHB, 2E:* Page 600.

Question #190. The SCBA unit and protective equipment add weight and bulk to the firefighter, causing increased exertion with loss of body fluids through perspiration. These actions increase during firefighting operations. Firefighters must be aware of them and of the symptoms of heat stress and their own limitations and abilities. During rehabilitation, EMS personnel should monitor vital signs and firefighters must hydrate to replace body fluids. NFPA 1001: 5.3.1. *FFHB, 2E:* Page 168, 169.

Question #191. NFPA 1001: 5.1.1.1. *FFHB, 2E:* Page 451.

Question #192. Methane has a vapor density of 0.6. If Methane escapes its container, it can be expected to rise into the air. NFPA 1001: 5.3; 6.3. *FFHB, 2E:* Page 87.

Question #193. Submersible pumps are very good tools for flooded basements. NFPA 1001: 5.3.14. *FFHB, 2E:* Page 635.

Question #194. The water motor gong, unless connected to another alarm, is a local alarm only. NFPA 1001: 6.5.1. *FFHB, 2E:* Page 331.

Question #195. Responders should understand the protective actions that are available to utilize. NFPA 472: 6.2.1 - 6.2.2. *FFHB, 2E:* Page 901.

Question #196. When Firefighter Hernandez is ready to transmit his message, he should first listen to the radio to make sure he does not interfere with other communications. When the airwaves are clear, he should depress the push-to-talk button and wait at least two seconds before beginning to speak to avoid clipping his message. NFPA 1001: 5.2.3. *FFHB, 2E:* Page 65-67.

Question #197. Pump-type extinguishers are hand-pumped devices of two designs, depending on whether the pump is internal or external to the tank. Pressurized-water, pressurized-loaded stream, and stored-pressure extinguishers operate by means of an expelling gas that propels the agent out of the container. NFPA 1001: 5.3.16. *FFHB, 2E:* Page 188.

Question #198. Hydraulics is the study of fluids at rest and in motion, which describes the flow pattern of water supply and fire streams. NFPA 1001: 5.3 NFPA 6.3. *FFHB, 2E:* Page 290.

Question #199. When transmitting messages on mobile and portable radios, firefighters should hold the microphone 1-2 inches from the mouth and at a 45-degree angle for a clear transmission. NFPA 1001: 5.2.3. *FFHB, 2E:* Page 65.

Question #200. Class A fires involve ordinary combustibles such as wood, paper, cloth, plastics, and rubber. NFPA 5.3.16. *FFHB, 2E:* Page 185.

Abandonment   Abandonment occurs when an emergency responder begins treatment of a patient and then leaves the patient or discontinues treatment prior to the arrival of an equally or higher trained responder.

Aboveground Storage Tank (AST)   Tank that is stored above the ground in a horizontal or vertical position. Smaller quantities of fuels are often stored in this fashion.

Abrasion   A scrape or brush of the skin usually making it reddish in color and resulting in minor capillary bleeding.

Absolute Pressure   The measurement of pressure, including atmospheric pressure. Measured in pounds per square inch absolute.

Absorption   A defensive method of controlling a spill by applying a material that absorbs the spilled chemical.

Accelerator   A device to speed the operation of the dry pipe valve by detecting the decrease in air pressure. It pipes air pressure below the clapper valve, speeding its opening.

Accident   The result of a series of events and conditions that lead to an unsafe situation resulting in injury and/or property damage.

Accident Chain   A series of events and conditions that can lead to or have led to an accident. These events and conditions are typically classified into five areas: environment, human factors, equipment, events, and injury.

Acclimation   The act of becoming accustomed or used to something. Typically achieved through repeated practice within a given set of conditions.

Acute   A quick one-time exposure to a chemical.

Adapter   Device that adapts or changes one type of hose thread to another, allowing connection of two different lines. Adapters have a male end on one side and a female on the other with each side being a different thread type, for example, an iron pipe to national standard adapter.

Administrative Warrant   An order issued by a magistrate that grants authority for fire personnel to enter private property for the purpose of conducting a fire prevention inspection.

Aerial Apparatus   Fire apparatus using mounted ladders and other devices for reaching areas beyond the length of ground ladders.

Air Bill   The term used to describe the shipping papers used in air transportation.

Aircraft Rescue and Firefighting (ARFF)   Of or pertaining to firefighting operations involving fixed or rotary wing aircraft.

Air Monitoring Devices   Used to determine oxygen, explosive, or toxic levels of gases in air.

Air-Purifying Respirators (APR)   Respiratory protection that filters contaminants out of the air, using filter cartridges. Requires the atmosphere to have sufficient oxygen, in addition to other regulatory requirements.

Allergic Reaction   The body's reaction to a substance to which there is an allergy.

Americans with Disabilities Act   Public law that bars discrimination on the basis of disability in state and local services. Enacted in 1990.

Amputation   Occurs when part of the body is severed completely as a result of an injury.

Anchor Point   A safe location from which to begin line construction on a wildland fire.

ANFO   The acronym that is used for ammonium nitrate fuel oil mixture, which is a common explosive. ANFO was used in the Oklahoma City bombing incident.

Anthrax   A biological material that is naturally occurring and is severely toxic to humans. It is commonly used in hoax incidents.

Application Rate   Amount of foam or foam solution needed to extinguish a fire. Usually expressed in gallons per minute per square foot or liters per minute per square meter.

Aqueous Film-Forming Foam (AFFF)   A synthetic foam that as it breaks down forms an aqueous layer or film over a flammable liquid.

Aquifer   A formation of permeable rock, gravel, or sand holding water or allowing water to flow through it.

Arson   A malicious fire or fires set intentionally by humans for vengeance or profit.

Arterial Bleeding   Bleeding from an artery.

Arteries   The blood vessels, or tubes, within the body that carry blood rich with oxygen and nutrients away from the heart.

Articulating Boom Ladder   An apparatus with a series of booms and a platform on the end. It is maneuvered into position by adjusting the various boom sections into place to position the platform at the desired location.

Aspect   The direction a slope faces given in compass directions.

Asphyxiation   Condition that causes death due to lack of oxygen and an excessive amount of carbon monoxide or other gases in the blood.

Association of Public Safety Communications Officials-Int., Inc. (APCO)   International not-for-profit organization dedicated to the advancement of public safety communications. Membership is made up of public safety professionals from around the world.

Atmospheric Pressure   The pressure exerted by the atmosphere, which for Earth is 14.7 pounds per square inch at sea level.

Atomization   The separation of atoms and molecules into an unconnected state where they are in suspension rather than in liquid form.

Attack Hose   Small- to large-diameter hose used to supply nozzles and other applicators or protective system for fire attack. Attack hose commonly means handheld hoselines from 1½ to 2½ inches (38 or 63 mm) in diameter.

Authority Having Jurisdiction (AHJ)   The responsible governing organization or body having legal jurisdiction.

Autoextended   When a fire goes out the window on one floor, up the side of the building, which is often noncombustible, and extends through the window or cockloft directly above.

Automated External Defibrillator (AED)   A portable computer-driven device that analyzes a patient's heart rhythm and delivers defibrillation shocks when necessary.

Automatic or **Constant Pressure Nozzle**   Nozzle with a spring mechanism built in that reacts to pressure changes and adjusts the flow and resultant reach of the nozzle.

Automatic Sprinkler System   A system of devices that will activate when exposed to fire, connected to a piping system that will supply water to control the fire. Typically, an automatic sprinkler system is also supported by firefighters when they arrive on the scene.

Auxiliary Appliances   Another term for protective devices, particularly sprinkler and standpipe systems.

Available Flow   Amount of water that can be moved to extinguish the fire. Depends on the water supply, pump(s) and their capabilities, and the size and length of hose.

Avulsion   An injury where a part of the skin is torn away, but still attached, leaving a flap or loose area hanging.

Awareness Level   The basic level of training for emergency response to a chemical accident, the basis of which is the ability to recognize a hazardous situation and call for assistance.

Axial Load   A load passing through the center of the mass of the supporting element, perpendicular to its cross section.

Backdraft   A sudden, violent reignition of the contents of a closed container fire that has consumed the oxygen within the space when a new source of oxygen is introduced.

Backflow Preventers   A check valve or set of valves used to prevent a backflow of water from one system into another. Required where a building water or fire protection system connects with the public water system. Backflow preventers are being required for environmental and health reasons.

Backstretch or **Flying Stretch**   An attack line lay where the engine is at the hydrant and the line is stretched back from the engine to the fire. The flying stretch is a version of the backstretch where the engine stops in front of the fire, the attack portion is removed, and the engine proceeds to the hydrant.

Balloon Frame   A style of wood frame construction in which studs are continuous for the full height of a building.

Bank Down   A condition in which the heat, smoke, and fire gases have reached the uppermost level in a compartment and, instead of continuing up, begin to push down from the ceiling toward the floor.

Base Radio   Radio station that contains all of the antennas, receivers, and transmitters necessary to transmit and receive messages.

Basic 9-1-1   Telephone system that automatically connects a person dialing the digits "9-1-1" to a predetermined answering point through normal telephone service facilities. Number and location information is not normally provided in *basic* systems.

Beam   A structural member subjected to loads perpendicular to its length.

Bed Ladder   The nonextending part of an extension ladder.

Bevel   The outside curve of the fork end of the Halligan tool.

Bight   A doubled section of rope, usually made along the standing part, that forms a U-turn in the rope that does not cross itself.

Biological Agents   Microorganisms that cause disease in humans, plants, and animals; they also cause the victim's health to deteriorate. Biological agents have been designed for warfare purposes.

Biomimetic   A form of gas sensor that is used to determine levels of carbon monoxide. It is of the type of sensors used in home CO detectors. It closely re-creates the body's reaction to CO and activates an alarm.

Blister Agents   A group of chemical agents that cause blistering and irritation of the skin. Sometimes referred to as vesicants.

Blood Agents   Chemicals that affect the body's ability to use oxygen. If they prevent the body from using oxygen, fatalities result.

Body Substance Isolation Precautions   A set of precautions for emergency responders designed to prevent exposure to any body fluid or substance.

Boiling Liquid Expanding Vapor Explosion (BLEVE)   Describes the rupture of a container when a confined liquid boils and creates a vapor pressure that exceeds the container's ability to hold it.

Boiling Point   The temperature at which liquids must be heated in order to turn into a gas.

Bolt Throw   The distance the bolt of a lock travels into the jamb or strike plate. Usually ½ to 1½ inches.

Bond   A substance or an agent that causes two or more objects or parts to bind.

Booster Hose   Smaller diameter, flexible hard-rubber-coated hose of ¾- or 1-inch (19- to 25-mm) size usually mounted on a reel that can be used for small trash and grass fires or overhaul operations after the fire is out.

Bourdon Gauge The type of gauge found on most fire apparatus that operates by pressure in a curved tube moving an indicating needle.

Box Canyon A canyon open on one end and closed on the other. They become very dangerous when wildfire enters them.

Brachial Artery A major artery in the inside of the upper arm that supplies blood to the arm. Can be used as a pressure point for controlling bleeding and for locating a pulse on an infant.

Branch The command designation established to maintain span of control over a number of divisions, sectors, or groups.

Bresnan Distributors Has six or nine solid tips or broken stream openings designed to rotate in a circular spray pattern. Used to fight fire in basements or cellars when firefighters cannot make a direct attack on the fire.

British Thermal Unit (BTU) A measurement of heat that describes the amount of heat required to raise 1 pound of water 1°F.

Brush Gear Another term for a wildland personal protective ensemble.

Building Officials Conference Association (BOCA) A group that establishes minimum building and fire safety standards.

Bulk Tank A large transportable tank, comparable to a tote, but considered to be the larger of the two.

Bump Test Used to determine if an air monitor is working. It will alarm if a toxic gas is present. It is a quick check to make sure the instrument responds to a sample of gas.

Bunkers A slang term that is used mostly to describe the components of a structural firefighting ensemble. The original use of the term *bunkers* referred only to the pant/boot combination that firefighters wore at night and placed next to their "bunks" for rapid donning.

Butyric Acid A fairly common lab acid that has been used in many attacks on abortion clinics. Although not extremely hazardous, it has a characteristic stench that permeates the entire area where it is spilled.

Bypass Eductor Eductor with two waterways and a valve that allows plain water to pass by the venturi or through the venturi to create foam solution.

Calibration Used to set the air monitor and to ensure that it reads correctly. When calibrating a monitor, it is exposed to a known quantity of gas to make sure it reads the values correctly.

Cantilever Beam A beam that is supported at only one end.

Capillaries The very small blood vessels in the body that connect arteries and veins and filter the oxygen and nutrients from the blood into the tissues of the body.

Capillary Bleeding Bleeding from a capillary.

Carbon Dioxide (CO₂) An inert colorless and odorless gas that is stored under pressure as a liquid that is capable of being self-expelled and is effective in smothering Class B and C fires.

Carbon Monoxide Colorless, odorless, poisonous gas that when inhaled combines with the red blood cells excluding oxygen.

Carcinogen A material that is capable of causing cancer in humans.

Cardiovascular System The heart, blood vessels, and blood within the body.

Carotid Pulse The pulse located on either side of the neck.

Catalytic Bead The most common type of combustible gas sensor that uses two heated beads of metal to determine the presence of flammable gases.

Ceiling Level The highest exposure a person can receive without suffering any ill effects. It is combined with the PEL, TLV, or REL as a maximum exposure.

Cellar Nozzles Has four spray nozzles designed to rotate in a circular spray pattern for fighting fires in basements or cellars when firefighters cannot make a direct attack on the fire.

Chain of Command Common fire service term that means to always work through one's direct supervisor. The fire service is viewed as a paramilitary organization and because of this all requests for information outside the assigned workplace should go through the supervisor.

Check Valves Valves installed to control water flow in one direction, typically when different systems are interconnected.

Chemical Burns Burns caused by chemical substances that come into contact with the skin or tissues of the body, creating a caustic reaction.

Chemtrec The Chemical Transportation Emergency Center, which provides technical assistance and guidance in the event of a chemical emergency; a network of chemical manufacturers who provide emergency information and response teams if necessary.

Chimney Another term for drainage. Given because of the draw of fire as in heat going up the chimney.

Chip Measuring System (CMS) A form of colorimetric air sampling in which the gas sample passes through a tube. If the correct color change occurs, the monitor interprets the amount of change and indicates a level of the gas on an LCD screen.

Choking Agents Agents that cause a person to cough and have difficulty breathing. The terrorism agents that are considered choking agents are chlorine and phosgene, both very toxic gases.

Chord The top and bottom components of a beam or truss. The top chord is subjected to compressive force; the bottom chord is subjected to tensile force.

Chronic A continual or repeated exposure to a hazardous material.

Cistern An underground water tank made from natural rock or concrete. Cisterns store large quantities of water—30,000 gallons or more—in areas without other water supplies or as a backup supply.

Clandestine Drug Labs Illegal labs set up to manufacture street drugs.

Class A Classification of fire involving ordinary combustibles such as wood, paper, cloth, plastics, and rubber.

Class B Classification of fire involving flammable and combustible liquids, gases, and greases. Common products are gasoline, oils, alcohol, propane, and cooking oils.

Class C Classification of fire involving energized electrical equipment, which eliminates using water-based agents.

Class D Classification of fire involving combustible metals and alloys such as magnesium, sodium, lithium, and potassium.

Class K A new classification of fire as of 1998 that involves fires in combustible cooking fuels such as vegetable or animal oils and fats.

Clipping Term associated with the use of two-way radios that is used to describe instances when either the first part of a message or the last part of a message is cut off as the result of either speaking before pressing the transmit key or releasing the transmit key prior to the end of a transmission.

Closed-Circuit SCBA A type of SCBA unit in which the exhaled air remains in the system to be filtered and mixed with oxygen for reuse.

Cockloft The area between the roof and the ceiling.

Code of Federal Regulations (CFR) The documents that include federally promulgated regulations for all federal agencies.

Collapse Zone The area around a building where debris will land when it falls. As an absolute minimum this distance must be at least 1½ times the height of the building.

Colorimetric Tubes Crystal-filled tubes that change colors in the presence of the intended gases. These tubes are made for the detection of known and unknown gases.

Column A structural element that is subjected to compressive forces—typically a vertical member.

Combination Attack A combined attack based on partial use of both offensive and defensive attack modes.

Combination Fire Attack A blend of the direct and indirect fire attack methods, with firefighters applying water to both the fuel and the atmosphere of the room.

Combination Nozzle A spray nozzle that is capable of providing straight stream and spray patterns, which are adjustable or variable by the operator. Most fog nozzles used today are combination nozzles.

Combustion The chemical action in which heat and light are produced and the heat is used to maintain the chemical chain reaction to continue the process.

Command Vehicle Typically used by operations chief officers in the fire service.

Common Terminology The designation of a term that is the same throughout an IMS.

Communicable Disease A disease that can be transmitted from one person to another.

Communications Sending, giving, or exchanging of information.

Company A team of firefighters with apparatus assigned to perform a specific function in a designated response area.

Compound A combination of substances joined in a chemical bond that exists in a proportional amount and cannot be separated without chemical interaction.

Compressed Air Foam System (CAFS) A foam system where compressed air is injected into the foam solution prior to entering any hoselines. The fluffy foam created needs no further aspiration of air by the nozzle.

Compression A force that tends to push materials together.

Computer-Aided Dispatch Computer-based automated system that assists the telecommunicator in assessing dispatch information and recommends responses.

Computer-Aided Management for Emergency Operations (CAMEO) Program A computer program that combines a chemical information database with emergency planning software. It is commonly used by HAZMAT teams to determine chemical information.

Concentrated Load A load applied to a small area.

Confined Space A space that is large enough to be entered but is not designed for continuous occupancy.

Conflagration A large and destructive fire.

Consent The acceptance of emergency medical treatment by a patient or victim.

Consist The shipping papers that list the cargo of a train. The listing is by railcar, and the consist lists all of the cars.

Consolidated Incident Action Plan The strategic goals to eliminate the hazard or control the incident.

Constant or **Set Volume Nozzle** Nozzle with one set volume at a set pressure. For example, 60 gpm at 100 psi (227 L/min 690 kPa). The only adjustment is the pattern.

Constricted A condition of the pupils where they are much smaller than normal and may appear almost like a "pinpoint."

Continuous Beam A beam that is supported in three or more places.

Control Room A room on the ground floor of a high-rise building where all building systems controls are located.

Cribbing The use of various dimensions of lumber arranged in systematic stacks (pyramid, box, step, etc.) to support an unstable load.

Critical Incident Stress Debriefing (CISD) A formal gathering of incident responders to help defuse and address stress from a given incident.

Critical Incident Stress Management (CISM) A process for managing the short- and long-term effects of critical incident stress reactions.

Cryogenic gas Any gas that exists as a liquid at a very cold temperature, always below −150°F.

Cutting Tools The group of tools used to cut through or around materials.

Damming The stopping of a body of water, which at the same time stops the spread of the spilled material.

Dangerous Cargo Manifest (DCM) The shipping papers for a ship, which lists the hazardous materials on board.

Database Organized collection of similar facts.

Dead Load The weight of the building materials and any part of the building permanently attached or built-in.

DECIDE Process A management system used to organize the response to a chemical incident. The factors of DECIDE are detect, estimate, choose, identify, do the best, and evaluate.

Decontamination The physical removal of contaminants (chemicals) from people, equipment, or the environment. Most often used to describe the process of cleaning to remove chemicals from a person.

Defensive Attack A calculated attack on part of a problem or situation in an effort to hold ground until sufficient resources are available to convert to an offensive form of attack.

Deflagrates Rapid burning, which in reality with regard to explosions can be considered a slow explosion, but is traveling at a lesser speed than a detonation.

Dehydration A loss of water and vital fluids in the body.

Deluge Systems Designed to protect areas that may have a fast-spreading fire engulfing the entire area. All of its sprinkler heads are already open, and the piping contains atmospheric air. When the system operates, water flows to all heads, allowing total coverage. The system uses a deluge valve that opens when a separate fire detection system senses the fire and signals to trip the valve open.

Density The mass per unit volume of a substance under specified conditions of pressure and temperature.

Deployment Plan Predetermined response plan of apparatus and personnel for specific types of incidents and specific locations.

Depth of Char A term commonly used by fire investigators to describe the amount of time wooden material had burned. The deeper the char, the longer the material was burning or exposed to direct flame.

Design Load A load the engineer planned for or anticipated in the structural design.

Detergent-Type Foams Use synthetic surfactants to break down the surface tension of water to create a foaming blanket.

Diffusion A naturally occurring event in which molecules travel from levels of high concentration to areas of low concentration.

Diking A defensive method of stopping a spill. A common dike is constructed of dirt or sand and is used to hold a spilled product. In some facilities, a dike may be preconstructed such as around a tank farm.

Dilated A condition of the pupils where they are much larger than normal and can take up almost the whole colored portion of the eye.

Dilution The addition of a material to the spilled material to make it less hazardous. In most cases water is used to dilute a spilled material, although other chemicals could be used.

Direct Fire Attack An attack on the fire made by aiming the flow of water directly at the material on fire.

Disassembly The actual taking apart of vehicle components.

Discharge Flow Total amount of water flowing from the discharge side of the pump.

Displacement The relocating of major parts (i.e., doors, roof, dash, steering column) of a vehicle.

Distortion The bending of sheet metal or components.

Distributed Load A load applied equally over a broad area.

Distributor Pipe or **Extension Pipe** Devices that allow a nozzle or other device to be directed into holes to reach basements, attic, and floors that cannot be accessed by personnel. The distributor pipe has self-supporting brackets that help hold it into place when in use.

Diverting Using materials to divert a spill around an item. For instance, several shovels full of dirt can be used to divert a running spill around a storm drain.

Division Command designation responsible for operations within an assigned geographic area.

Double Female Allows the two male ends of hose to be connected.

Double Male Used to connect two female thread couplings.

Drafting The pumping of water from a static source by taking advantage of atmospheric pressure to force water from the source into the pump.

Drainage A topographic feature on the side of a hill or mountain that naturally collects water runoff, channeling it to the bottom of the rise. Fire is attracted to this feature.

Dressing The practice of making sure that all parts of a knot are lying in the proper orientation to the other parts and look exactly as the pictures herein indicate.

Dry Chemicals Dry extinguishing agents divided into two categories. Regular dry chemicals work on Class B and C fires; multipurpose dry chemicals work on Class A, B, and C fires.

Dry Hydrant   A piping system for drafting from a static water source with a fire department connection at one end and a strainer at the water end.

Dry Pipe Systems   Air under pressure replaces the water in the system to protect against freezing temperatures. The sprinkler control valve uses a dry pipe valve to keep pressurized air maintained above with the supply water under pressure below the valve.

Dry Powders   Extinguishing agents for Class D fires.

Dump Site   The area where tenders are unloaded or their load dumped.

Dutchman   A short fold of hose or a reverse fold that is used when loading hose and a coupling comes at a point where a fold should take place or when two sets of couplings end up on top of or next to each other. The dutchman moves the coupling to another point in the load.

Dynamic   A rope having a high degree of elongation (10 to 15 percent) at normal safe working loads.

Ears   Elongated folds or flaps at the ends of a layer of hose to assist in pulling that layer.

Eccentric Load   A load perpendicular to the cross section of the supporting element that does not pass through the center of mass.

Eductor   Device that siphons a liquid from a container into a moving stream.

8-Step Process   A management system used to organize the response to a chemical incident. The elements are site management and control, identifying the problem, hazard and risk evaluation, selecting PPE and equipment, information management and resource coordination, implementing response objectives, decon and cleanup operations, and terminating the incident.

Electrical Conductor   Any material that will permit electricity to flow through it.

Emergency Call Box   System of telephones connected by private line telephone, radio-frequency, or cellular technology usually located in remote areas and used to report emergency situations.

Emergency Communications Center   Facility either wholly or partially dedicated to being able to receive emergency and, in some instances, nonemergency reports from citizens. Centers such as these are sometimes referred to as fire alarm, headquarters, dispatch, or a public safety answering point (PSAP).

Emergency Decon   The rapid removal of a material from a person when that person (or responder) has become contaminated and needs immediate cleaning. Most emergency decon setups use a single hoseline to perform a quick gross decon of a person with water.

Emergency Medical Dispatch   System designed for use by telecommunicators to assist them in evaluating patient symptoms using predetermined criteria and responses.

Emergency Medical Services   The delivery of prehospital medical treatment.

Emergency Medical Technician (EMT)   An individual trained and certified to provide basic life support emergency medical care.

Emergency Planning and Community Right to Know Act (EPCRA)   The portion of SARA that specifically outlines how industries report their chemical inventory to the community.

Emergency Response Guidebook (ERG)   Book provided by the DOT that assists the first responder in making decisions at a transportation-related chemical incident.

Emergency Response Planning (ERP)   Levels that are used for planning purposes and are usually associated with the preplanning for evacuation zones.

Employee Assistance Program (EAP)   A defined program that offers professional mental health and other health services to employees.

Encapsulated Suit   A chemical suit that covers the responder, including the breathing apparatus. Usually associated with Level A clothing, that is gas- and liquid tight, but there are some Level B styles that are fully encapsulated, but not gas- or liquid tight.

Encoder   Device that converts an "entered" code into paging codes, which in turn activate a variety of paging devices.

Endothermic Reaction   A chemical reaction in which heat is absorbed, and the resulting mixture is cold.

Engine Company   The unit designation of a group of firefighters assigned to a piece of apparatus designed to deliver water to the fire scene.

Engulfed   To swallow up or overwhelm.

Enhanced 9-1-1   Similar in nature to basic 9-1-1 but with the capability to provide the caller's telephone number and address.

Equilibrium   When referring to gas or liquids, a state where a balance has occurred in mixture or weight.

Etiological   A form of a hazard that includes biological, viral, and other disease-causing materials.

Evacuation   The movement of people from an area, usually their homes, to another area that is considered to be safe. People are evacuated when they are no longer safe in their current area.

Evaporation   A process in which the molecules of a liquid are liberated into the atmosphere at a rate greater than the rate at which the molecules return to the liquid. Ultimately the liquid becomes fully airborne in a gaseous state.

Exhauster   A device to speed the operation of the dry pipe valve by detecting the decrease in air pressure. It helps bleed off air.

Exit Drills in the Home (EDITH)   A fire survival program to encourage people to practice fire drills from their home or residence.

Exothermic Reaction   A chemical reaction that releases heat, such as when two chemicals are mixed and the resulting mixture is hot.

Explosive Limits   A concentration of a gas or liquid that is not too rich or too lean to ignite with force.

Exposure   A contact with a potentially disease-producing organism; the contact does not necessarily produce the disease in the exposed individual.

Exposure Fire   Any combustible item threatened by something burning nearby that has caught on fire.

Extension Ladder   A ladder consisting of two or more sections that has the ability to be extended to a desired height through the use of a halyard.

External Bleeding   Bleeding that is coming from an open wound on the body.

External Floating Roof Tank   Tank with the roof exposed on the outside that covers the liquid within the tank. The roof floats on the top of the liquid, which does not allow for vapors to build up.

Extremely Hazardous Substances (EHS)   A list of 366 substances that the EPA has determined present an extreme risk to the community if released.

Extricate   To set free, release, or disentangle a patient from an entrapment situation.

Federal Communications Commission   Government agency charged with administering the provisions of the Communications Act of 1934 and the revised Telecommunications Act of 1996 and is responsible for nonfederal radio-frequency users.

Femoral Artery   A major artery in the lower body near the groin that supplies the leg with blood. Can be used as a pressure point for controlling bleeding in the lower extremities.

Fill Site   The area where tenders are filled or get their water.

Fine Decon   The most detailed of the types of decontamination. Usually performed at a hospital that has trained staff and is equipped to perform fine decon procedures.

Fire Alarm   Notification to the fire department that a fire or other related emergency is in progess, which results in a response.

Fire Engineering   The study of fire, fire behavior, fire extinguishment, and suppression.

Fire Flow Capacity   The amount of water available or amount that the water distribution system is capable of flowing.

Fire Flow Requirement   A measure comparing the amount of heat the fire is capable of generating versus the amount of water required for cooling the fuels below their ignition temperature.

Fire Hazard   Any condition, situation, or operation that could lead to the ignition of unwanted combustion or result in proper combustion becoming uncontrolled.

Fire Hose   A flexible conduit used to convey water or other agent from a water source to the fire.

Fire Hydraulics   The principles associated with the storage and transfer of water in firefighting activities.

Fire Intensity   A measurement of Btus produced by a fire. Sometimes measured in flame length in the wildland environment.

Fire Load   The amount of heat energy released when combustibles burn in a given area or building— expressed in British thermal units (Btus).

Fire Resistive   The capacity of a material to withstand the effects of fire.

Fire-Resistive Rating   The time in hours that a material or assembly can withstand fire exposure. Fire-resistive ratings are usually provided for testing organizations. The ratings are expressed in a time frame, usually hours or portions thereof.

Fire Shelter   A last-resort protective device for wildland firefighters caught or trapped in an environment where a firestorm or blowup is imminent.

Fire Societies   Groups of people who voluntarily banded together to deal with a community's fire problems.

Fire Station Alerting System   System used to transmit emergency response information to fire station personnel via voice and/or digital transmissions.

Fire Stopping   Pieces of material, usually wood or masonry, placed in stud or joist channels to slow the extension of fire.

Fire Stream   The water or other agent as it leaves the hose and nozzle toward its objective, usually the fire.

Fire Tetrahedron   Four-sided pyramid-like figure describing the heat, fuel, oxygen, and chemical reaction necessary for combustion.

Fire Wardens   Designated community individuals who walked the streets at night looking for fire and carrying large wooden rattles with which to signify a found fire.

Fire Watch   An organized patrol of a protected property when the sprinkler or other protection system is down for maintenance. Personnel from the property regularly check to make sure a fire has not started and assist in evacuation and prompt notification of the fire department.

Firefighter Assist and Search Team (FAST)   A company designated to search for and rescue trapped or lost firefighters. May also be called a rapid intervention team (RIT).

Firemark   Signs on sheets of metal telling firefighters which company held the insurance policy on a home or building.

First Responders   A group designated by the community as those who may be the first to arrive at a chemical incident. This group is usually composed of police officers, EMS providers, and firefighters.

Fit Testing   A test that ensures the respiratory protection fits the face and offers maximum protection.

Flammable Limits   The concentration level of a substance at which it will burn.

Flammable Range   Ratio of gas to air that will sustain fire if exposed to flame or spark.

Flanks of the Fire   The sides of a wildland fire running from the start point up each side to the end of the fire running into unburned areas.

Flash Point   The temperature at which a liquid will liberate a flammable gas.

Flashover   A sudden event that occurs when all the contents of a container reach their ignition temperature simultaneously.

Flow   The rate or quantity of water delivered, usually measured in gallons per minute or liters per minute (1 gpm = 3.785 L/min).

Fluoroprotein Film–Forming Foam (FFFP)   Combines protein with the film-forming fluorinated surfactants of AFFF to improve on the qualities of both types of foam.

Fluoroprotein Foam   Designed as an improved protein foam with a fluorinated surfactant added.

Flush or **Slab Doors**   Doors that are flat or have a smooth surface and may be of either hollow-core or solid-core construction.

Fly Ladder   That portion of a ladder that extends out from the bed ladder. Also called *fly section.*

Foam   An aggregate of gas-filled bubbles formed from aqueous solutions of specially formulated concentrated liquid foaming agents.

Fog Nozzle   Delivers either a fixed spray pattern or variable combination of straight stream and spray patterns.

Forcible Entry   The fire scene task of gaining entry to a building or secured area by disabling, breaking, or going around locking and security devices.

Foreman   Individual designated as the leader of an early fire company; a predecessor to the modern title of fire chief.

Forestry Hose   Specially designed hose for use in forestry and wildland firefighting. It comes in 1- and 1½-inch (25- and 38-mm) sizes and should meet U.S. Forestry Service specifications.

Formal Decon   The washing and scrubbing portion of the decontamination process. The process is usually repeated and is performed by a decon crew.

Fracture   A medical term for a broken or cracked bone in the body.

Frangible Disk   A type of pressure-relieving device that actually ruptures in order to vent the excess pressure. Once opened the disk remains open; it does not close after the pressure is released.

Freelancing   The act of working alone or performing a task for which the firefighter has not been assigned.

Freezing point   The temperature at which liquids become solids.

Friction   Caused by the rubbing of materials against each other while in movement and converts or robs some of the movement energy into heat energy.

Friction Loss   Measurement of friction in a system such as a hoseline.

Frontage   The portion of a property that faces and actually touches the street.

Fuel Resistance   Ability to tolerate the fuel and to avoid being saturated by or picking up the fuel.

Full Thickness Burns   Burns affecting not only the skin structure but the tissues and muscles underneath. Full thickness burns may be red, white, or charred in color, and will appear dry because the blood vessels in the skin are damaged extensively and are not supplying fluids to the area.

Garden Apartment   A two- or three-story apartment building with common entryways and layouts on each floor, surrounded by greenery and landscaping, sometimes having porches and patios.

Gas   A state of matter that describes the material in a form that moves freely about and is difficult to control. Steam is an example.

Gate Valves   Indicating and nonindicating valves that are opened and closed to control water flow.

Gauge Pressure   Measures pressure without atmospheric pressure. Normally fire department gauges do not measure atmospheric pressure. Gauge pressure is measured in psi or psig.

GEDAPER Process   A management system used to organize the response to a chemical incident. The factors are gather information, estimate potential, determine goals, assess tactical options, plan, evaluate, and review.

Girder   A large structural member used to support beams or joists—that is, a beam that supports beams.

Glazing   The glass or other clear material portion of the window that allows light to enter.

Gross Decon   The portion of the decontamination process that removes the majority of the chemicals through a flushing process. The gross washing is done using large amounts of water and is usually done by the individual or the individual's partner.

Gross Negligence   Occurs when an individual disregards training and continues to act in a manner without regard for others.

Ground Pads   Sheets of plywood, planks, aluminum sheets, and so on, used to distribute weight over a larger area.

Guard Dogs   Trained animals that will attack an intruder.

Guideline/Lifeline   Rope used as a crew is searching a structure to assist them in finding their way back out.

Gusset Plate   A connecting plate used in truss construction. In steel trusses, these plates are flat steel stock. In wood trusses, the plates are either light-gauge metal or plywood.

Halligan Tool   From the prying group, a 30-inch forged steel tool with three primary parts: the adz end, the pike end, and the fork end.

Halyard   A rope or cable that is used to raise the fly ladders of an extension ladder.

Hard Suction Hose A special type of hose that does not collapse when used for drafting.

Hardware Equipment used in conjunction with life safety ropes and harnesses (carabiners, figure eights, rappel racks, etc.).

Harnesses Webbing sewn together to form a belt, seat harness, or seat and chest harness combination.

Hazardous Materials Chemicals that are flammable, explosive, or otherwise capable of causing death or destruction when improperly handled or released.

Hazardous Materials Technician An individual trained to meet the requirements of CFR OSHA 1910.120, *Technician Level for Hazardous Materials Response.*

Hazardous Waste Operations and Emergency Response (HAZWOPER) The OSHA regulation that covers safety and health issues at hazardous waste sites, as well as response to chemical incidents.

HAZMAT Crime A criminal act that uses or threatens the use of chemicals as a weapon.

Head of the Fire The running top or aggressive end of the fire away from the start point.

Head Pressure Measures the pressure of a column of water in feet (meters). Head pressure gain or loss results when water is being pumped above or below the level of the pump. A head of 2.31 feet (0.7 m) would equal 1 psi (6.895 kPa).

Heat Resistance The ability of foam to stand up to the heat of the fire or to hot surfaces near the fire.

Heat Sink The term used to denote a place where heat is drained away from a source.

Helix The metal or plastic bands or rings used in hard suction hose to prevent its collapse under drafting conditions.

Higbee Cut The blunt ending of the threads of fire hose couplings that allows the threads to be properly matched, avoiding cross-threading.

Hoistway The shaft in which an elevator or a number of elevators travel.

Hollow-Core Door Any door that is not solid, usually with some type of filler material between face panels.

Home Alerting Devices Emergency alerting devices primarily used by volunteer department personnel to receive reports of emergency incidents.

Hook A tool with a 32-inch to 12-foot handle with a pike and hook on one end. Used for pulling ceilings or separating other materials. Also known as a *pike pole.*

Horizontal Ventilation Channeled pathway for fire ventilation via horizontal openings.

Hose Bed The portion or compartment of fire apparatus that carries the hose.

Hose Bridges Devices that allow vehicles to pass over a section of hose without damaging it.

Hose Cap Does not allow water to flow through it. Instead, it caps the end of a hoseline or appliance to prevent water flow.

Hose Cart A handcart or flat cart modified to be able to carry hose and other equipment around large buildings. Some departments use them for high-rise situations.

Hose Clamp A device to control the flow of water by squeezing or clamping the hose shut. Some work by pushing a lever that closes the jaws of the device and others have a screw mechanism or hydraulic pump that closes the jaws.

Hose Jackets Metal or leather devices used for stopping leaks without shutting down the line that is fitted over the leaking area and either clamped or strapped together to control the leak.

Hose Roller or **Hoist** A metal frame, with a securing rope, shaped to fit over a windowsill or edge of a roof with two rollers to allow the hose to roll over the edge, preventing chafe.

Hose Strap A short strap with a forged handle and cinch clip attached. Used to help maneuver hose and attach hose to ladders and stair rails.

HVAC Acronym for heating, ventilation, and air-conditioning unit. HVACs are typically a rooftop unit on commercial buildings. Buildings may have one or dozens of these units.

Hydrant Valves or **Switch Valves** Valve used on a hydrant that allows an engine to connect and charge its supply line immediately but also allows an additional engine to connect to the same hydrant without shutting down the hydrant, and increases the flow of the hydrant.

Hydrant Wrenches Tools used to operate the valves on a hydrant. May also be used as a spanner wrench. Some are plain wrenches and others have a ratchet feature to speed the operation of the valve.

Hydraulic Pistons Mechanical rams that operate by pressure exerted through the use of a liquid, usually some form of oil.

Hydraulics The study of fluids at rest and in motion.

Hydrocarbon Any of numerous organic compounds, such as benzene and methane, that contain only carbon and hydrogen.

Hyperbaric Chamber A chamber that is usually used to treat scuba divers who ascended too quickly and need extra oxygen to survive. The chamber re-creates the high-pressure atmosphere of diving and forces oxygen into the body. It is also successful in the treating of carbon monoxide poisoning and smoke inhalation, because both of these problems require high amounts of oxygen to assist with the patient's recovery.

Hypoperfusion A serious condition caused by a problem or failure of the circulatory system that results in a decrease of oxygen and vital nutrients to the body's tissues. Also known as shock.

Hypoxia A deficiency of oxygen.

ICt$_{50}$ The incapacitating level for time to 50 percent of the exposed group. It is a military term that is often used in conjunction with LCt$_{50}$.

Ignition   The point at which the need for outside heat application ceases and a material sustains combustion based on its own generation of heat.

Ignition Point   The temperature at which a substance will continue to burn after the source is removed.

Ignition Temperature   The temperature of a liquid at which it will ignite on its own without an ignition source. Can be compared to SADT.

Immediately Dangerous to Life and Health (IDLH)   The maximum level of danger one could be exposed to and still escape without experiencing any effects that may impair escape or cause irreversible health effects.

Impact Load   A load that is in motion when it is applied.

Implied Consent   The assumption of acceptance of emergency medical treatment by an unconscious patient or a child with no parents or legal guardians present.

Incendiary Agents   Chemicals that are used to start fires, the most common being a Molotov cocktail.

Incident Action Plan (IAP)   A strategic and tactical plan developed by the incident commander.

Incident Commander Level   A training level that encompasses the operations level with the addition of incident command training. Intended to be the person who may command a chemical incident.

Incident Management System (IMS)   A management system utilized on the emergency scene that is designed to keep order and follow a sequence of set guidelines.

Incision   A cut to the skin that leaves a straight, even pattern.

Increaser   Used to connect a smaller hose to a larger one.

Indirect Fire Attack   An attack made on interior fires by applying a fog stream into a closed room or compartment, thus converting the water into steam to extinguish the fire.

Infection Control   Procedures and practices for firefighters and emergency medical care providers to follow to prevent the transmission of diseases and germs from a patient to themselves or other patients.

Infectious Disease   See **Communicable Disease.**

Infrared Sensor   A sensor that uses infrared light to determine the presence of flammable gases. The light is emitted in the sensor housing and the gas passes through the light. If it is flammable the sensor will indicate the presence of the gas.

Initial Assessment   The initial investigative action taken by care providers to determine if the patient has the basic signs of life as well as any serious, life-threatening injuries.

In-Line Eductor   Eductor in which the waterway is always piped through a venturi.

Inorganic   A substance that is not of any living organism.

Intake Relief Valve   Required on large-diameter hose at the receiving engine that functions as a combined overpressurization relief valve, a gate valve, and an air bleed-off.

Integrated Communications   The ability of all units or agencies to communicate at an incident.

Interface Firefighting   Fighting wildland fire and protecting exposed structures in rural settings.

Intermodal Containers   These are constructed in a fashion so that they can be transported by highway, rail, or ship. Intermodal containers exist for solids, liquids, and gases.

Internal Bleeding   Bleeding within the body when no visible open wound is present.

Internal Floating Roof Tank   Tank with a roof that floats on the surface of the stored liquid, but also has a cover on top of the tank, so as to protect the top of the floating roof.

Intervention   The act of intervening; to come between as an influencing force. Typically a reactive action.

Irons   The combination of a Halligan tool and flathead ax or maul.

Irritant   A material that is irritating to humans, but usually does not cause any long-term adverse health effects.

Isolation Area   An area that is set up by responders and is intended to keep people, both citizens and responders, out. May later become the hot zone/sector as the incident evolves. Is the minimum area that should be established at any chemical spill.

Jacket   The outer part of the hose, often a woven cloth or rubberized material, which protects the hose from mechanical and other damage.

Jamb   The mounting frame for a door.

Jet Dump   A device that speeds the process of dumping a load of water from a tanker/tender.

Jet Siphon   A device that speeds the process of transferring water from one tank to another.

Joist   A wood framing member that supports floor or roof decking.

Kerf Cut   A quick and easily made examination hole. It is created by letting the spinning blade of a power saw cut through the material to be cut and pulling it out, leaving only a slit-like cut measuring approximately 12 inches long and only as wide as the cutting blade.

Kern   A derivative of the term *kernel,* which is defined as "the central, most important part of something; core; essence."

Knockdown Speed   Speed with which foam spreads across the surface of a fuel.

Laceration   A cut to the skin and underlying tissues that leaves an irregular, even pattern.

Ladder Pipe   An appliance that is attached to the underside of an aerial ladder for an elevated water application.

Laminated Glass   Glass composed of two or more sheets of glass with a plastic sheet between them. The purpose of the plastic sheet is to hold the glass together if broken, thus reducing the hazard of flying glass.

Landing Plate   The plate at the top or bottom of an escalator where the steps disappear into the floor.

Laws   Legislation that is passed by the House and Senate and signed by the president.

LCt$_{50}$ The lethal concentration for time to 50 percent of the group. Same as the LC$_{50}$, but adds the element of time. It is a military term.

Leaking Underground Storage Tank (LUST) Describes a leaking tank that is underground.

Ledge Door   Door built with solid material, usually individual boards, common in barns and warehouses.

Lethal Concentration (LC$_{50}$)   A value for gases that provides the amount of chemical that could kill 50 percent of the exposed group.

Lethal Dose (LD$_{50}$)   A value for solids and liquids that provides the amount of a chemical that could kill 50 percent of an exposed group.

Level A Protective Clothing   Fully encapsulated chemical protective clothing. It is gas and liquid tight and offers protection against chemical attack.

Level B Protective Clothing   A level of protective clothing that is usually associated with splash protection. Level B requires the use of SCBA. Various clothing styles are considered Level B.

Liability   The possibility of being held responsible for individual actions.

Life Safety   Term applied to the fire protection concept in which buildings are designed to allow for the escape of building occupants without injuries. Life safety usually makes the building more fire resistant, but this is not the main goal.

Life Safety Line   According to NFPA 1983, rope dedicated solely to the purpose of constructing lines for supporting people during rescue, firefighting, or other emergency operations, or during training evolutions.

Lifting   A term used to describe the removal of upper-level smoke and heat when cool air replaces the upper-level hot air that is escaping.

Liner   The inner layer of fire hose, usually made of rubber or a plastic material, that keeps the water in the tubing of the hose.

Lintel   A beam that spans an opening in a load-bearing masonry wall.

Liquid   A state of matter that implies fluidity, which means a material has the ability to move as water would. There are varying states of being a liquid from moving very quickly to moving very slowly. Water is an example.

Live Load   The weight of all materials and people associated with but not part of a structure.

Load-Bearing Wall   Any wall that supports other walls, floors, or roofs.

Loaded Stream   Combats the water freezing problem by adding an alkali salt as an antifreezing agent.

Loading   The weight of building materials or objects in a building.

Local Application System   Designed to protect only a certain or local portion of the building, usually directly where the hazard will occur or spread.

Local Emergency Planning Committee (LEPC)   A group composed of members of the community, industry, and emergency responders to plan for a chemical incident, and to ensure that local resources are adequate to handle an incident.

Locking Devices   A mechanical device or mechanism used to secure a door or window.

Loop   A turn in the standing part that crosses itself and results in the standing part continuing on in the original direction of travel.

Lower Explosive Limit (LEL)   The lower part of the flammable range, and is the minimum required to have a fire or explosion.

Mantle   Anything that cloaks, envelops, covers, or conceals.

Mask Confidence or "Smoke Divers" Training Training courses designed to develop a firefighter's skills and confidence for using SCBA.

Mass Casualty   An incident in which the number of patients exceeds the capability of the EMS to manage the incident effectively. In some jurisdictions this can be two patients, while in others it may take ten to make the incident a mass casualty.

Master Stream or **Heavy Appliances**   Non-handheld water applicator capable of flowing over 350 gallons of water per minute (1325 L/min).

Mastery   The concept that an individual can achieve 90 percent of an objective 90 percent of the time.

Material Safety Data Sheet (MSDS)   Information sheet for employees that provides specific information about a chemical, with attention to health effects, handling, and emergency procedures.

Matter   Something that occupies space and can be perceived by one or more senses; a physical body, a physical substance, or the universe as a whole. Something that has mass and exists as a solid, liquid, or gas.

Mayday   A universal call for help. A Mayday indicates that an individual or team is in extreme danger.

Means of Egress   A safe and continuous path of travel from any point in a structure leading to a public way. Composed of three parts: the exit access, the exit, and the exit discharge.

Medium-Diameter Hose (MDH)   Either 2½- or 3-inch (63- or 75-mm) hose.

Medi-Vac An ambulance that transports patients by air. Typically, medi-vac units are helicopters with highly trained EMS personnel and nurses.

Melting Point The temperature at which solids become liquids.

Metal Oxide Sensor (MOS) A coiled piece of wire that is heated to determine the presence of flammable gases.

Midslope An area partway up a slope. Any location not on the bottom or top of a slope, as in a midslope road crossing the slope horizontally.

Miscible Having the ability to mix with water.

Mission Statement A written declaration by a fire agency describing the things that it intends to do to protect its citizenry or customers.

Mission Vision A term used to describe a condition in which a person becomes so focused on an objective that peripheral conditions are not noticed, as if the person is wearing blinders.

Mitigation Actions taken to eliminate a hazard or make a hazard less severe or less likely to cause harm. Typically a proactive action.

Mobile Data Computer Communications device that, unlike the mobile data terminal, does have information processing capabilities.

Mobile Data Terminal Communications device that in most cases has no information processing capabilities.

Mobile Radio Complete receiver/transmitter unit that is designed for use in a vehicle.

Mobile Support Vehicle Vehicle designed exclusively for use as an on-scene communication center and command post.

Modular Organization The ability to start small and expand if an incident becomes more complex.

Molecule The smallest particle into which an element or a compound can be divided without changing its chemical and physical properties; a group of like or different atoms held together by chemical forces.

Mortar Mixture of sand, lime, and portland cement used as a bonding material in masonry construction.

Mounting Hardware Hinges, tracks, or other means of attaching a door to the frame or jamb.

Multigas Detector A term used to describe an air monitor that measures oxygen levels, explosive (flammable) levels, and one or two toxic gases such as carbon monoxide or hydrogen sulfide.

Multiple-Alarm Incident Involves the response of additional personnel.

Mutual Aid or **Assistance Agreements** Prearranged written agreements of the type and amount of assistance one jurisdiction will provide to another in the event of a large-scale fire or disaster. The key to understanding mutual aid is that it is a reciprocal agreement.

National Emergency Number Association Not-for-profit organization founded in 1982 and made up of more than 6,000 members. The association fosters technical advancement, availability, and implementation of a universal emergency telephone number system.

National Fire Protection Association (NFPA) A not-for-profit membership organization that uses a consensus process to develop model fire prevention codes and firefighting training standards.

National Institute for Occupational Safety and Health (NIOSH) A federal institute tasked with investigating firefighter fatalities and making recommendations to prevent reoccurrence.

National Response Center (NRC) The location that must be called to report a spill if it is in excess of the reportable quantity.

Needed or **Required Flow** Estimate of the amount of water required to extinguish a fire in a certain type period. Based on the type and amount of fuel burning.

Negligence Acting in an irresponsible manner or different from the way in which someone was trained; that is, differing from the standard of care.

Nerve Agents Chemicals that are designed to kill humans, specifically in warfare. They are chemically similar to organophosphorus pesticides and cause the same medical reaction in humans.

Nested The state when all the ladders of an extension ladder are unextended.

NFPA 1001 *Standard for Fire Fighter Professional Qualifications,* a national consensus training standard establishing the job performance requirements of tasks to be performed by firefighters.

NFPA 1404 National Fire Protection Association standard created by the Fire Service Training Committee detailing the requirements for fire service SCBA programs, including training and maintenance procedures.

NFPA 1500 National Fire Protection Association standard created by the Technical Committee on Fire Service Occupational Safety and Health that addresses a number of issues concerning protective equipment.

NFPA 1981 National Fire Protection Association standard specific to open-circuit SCBA for fire service use that contains additional requirements above the NIOSH certification.

NFPA 72 National Fire Alarm Code.

NFPA Standard 1931 The standard issued by the National Fire Protection Association that governs fire service ladder testing and certification.

9-1-1 Emergency telephone number that provides access to the public safety services in the community, region, and, ultimately, nation.

NIOSH National Institute for Occupational Safety and Health, 42 CFR Part 84, sole responsibility for testing and certification of respiratory protection including fire service SCBA.

No-Knowledge Hardware Locking devices that require no key or special knowledge to operate.

Nozzle A tapered or constricted tube used to increase the speed or change the direction of water or other fluids.

Nozzle Flow The amount or volume of water that a nozzle will provide. Flow is measured in gallons per minute or liters per minute.

Nozzle Pressure The pressure required to effectively operate a nozzle. Pressure is measured in pounds per square inch or kilopascals.

Nozzle Reach The distance the water will travel after leaving the nozzle. Reach is a function of the pressure, which is converted to velocity or speed of the water leaving the nozzle.

Nozzle Reaction The force of nature that makes the nozzle move in the opposite direction of the water flow. The nozzle operator must counteract the thrust exerted by the nozzle to maintain control.

Occupancy Classifications The use for which a building or structure is designed.

Occupant Use Hose Hose that is used in standpipe systems for building occupants to fight incipient fires. It is usually 1½-inch (38-mm) single-jacket hose similar to attack hose.

Occupational Safety and Health Administration (OSHA) The federal agency, under the Department of Labor, that is responsible for employee occupational safety.

Offensive Attack An aggressive attack on a situation where resources are adequate and capable of handling the situation.

One- or Two-Person Rope According to NFPA 1983, a one-person rope requires a minimum tensile strength of 4,500 pounds, and a two-person rope requires a minimum tensile strength of 9,000 pounds.

Open-Circuit SCBA A type of SCBA unit in which the exhaled air is vented to the outside atmosphere.

Operational Period The time frames for operations at an incident. At large-scale or complex incidents these will usually be eight- to twelve-hour time frames.

Operations Level The next level of training above awareness that provides the foundation which allows for the responder to perform defensive activities at a chemical incident.

Ordinary Tank A horizontal or vertical tank that usually contains combustible or other less hazardous chemicals. Flammable materials and other hazardous chemicals may be stored in smaller quantities in these types of tanks.

Organic A substance derived from living organisms.

OSHA 29 CFR 1910.134 Standard establishing minimum medical, training, and equipment levels for respiratory protection programs.

Outside Stem and Yoke (OS&Y) Valve Has a wheel on a stem housed in a yoke or housing. When the stem is exposed or outside, the valve is open. Also called an outside screw and yoke valve.

Overpacked A response action that involves the placing of a leaking drum (or container) into another drum. There are drums made specifically to be used as overpack drums in that they are oversized to handle a normal size drum.

Oxidizer A catalyst in the breakdown of molecules.

Oxygen Deficient Atmosphere An atmosphere with an oxygen content below 19.5 percent by volume.

Packaging The bandaging and preparing of a patient to be moved from the place of injury to a stretcher.

Panel Doors Doors with a solid stile and rails with panels made of wood or glass or other materials.

Panic Hardware Hardware mounted on doors that enable them to be opened by pushing from the inside.

Paragraph q The paragraph within HAZWOPER that outlines the regulations that govern emergency response to chemical incidents.

Paramedic (EMT-P) An individual trained and certified to provide advanced life support emergency medical care, including drug therapy.

Parapet The projection of a wall above the roofline of a building.

Partial Thickness Burns Burns affecting the entire skin structure that lies over the top of the fatty tissues and muscles causing skin to turn red and blistering of the skin.

Passport A term given to a specific accountability system where crews are tracked using a card (passport) with all members listed. An accountability manager tracks the passports on an accountability board.

PDD 39 Presidential Decision Directive 39, which established the FBI as the lead agency in terrorism incidents responsible for crisis management. It also established FEMA as the lead for consequence management.

Permeation The movement of chemicals through chemical protective clothing on a molecular level; does not cause visual damage to the clothing.

Permissible Exposure Limit (PEL) An OSHA value that regulates the amount of a chemical that a person can be exposed to during an eight-hour day.

Personal Alert Safety System (PASS) A device that emits a loud alert or warning that the wearer is motionless.

Personal Size-Up A continuous mental evaluation of an individual's immediate environment, facts, and probabilities.

Personnel Accountability Report (PAR) This is an organized roll call of all units assigned to an incident.

Photo-ionization Detector (PID) An air monitoring device used by HAZMAT teams to determine the amount of toxic materials in the air.

Piercing Nozzles Originally designed to penetrate the skin of aircraft and now have been modified to pierce through building walls and floors.

Pike Pole   See **Hook.**

Pipe Chases   A construction term used to describe voids designed to house building water supply and waste pipes. The term *electrical chase* is used for wiring.

Pitot Gauge   A device with an opening in its blade-shaped section that allows water to flow to a Bourdon gauge and registers the flowing discharge pressure of an orifice.

Plan View   A drawing or diagram of a building or area as seen from directly overhead. May include a site plan or a floor plan.

Platform Framing   A style of wood frame construction in which each story is built on a platform, providing fire stopping at each level.

Polar Solvent   A material that will mix with water, diluting itself.

Polar Solvent Type of Foam or Alcohol-Resistant Foam   Foam that is compatible with alcohol and/or polar solvents by creating a polymeric barrier between the water in the foam and the polar solvent.

Polymeric Barrier   A separation barrier made up of polymer or a chain of molecules linked in a series of long strands. This separates a polar solvent from an ATC foam blanket.

Polymerize   A chain reaction in which the material quickly duplicates itself and, if contained, can be very explosive.

Portable Hydrant or Manifold   Like a large water thief and may have one or more intakes and numerous outlets to allow multiple hoselines to be utilized with or without a pumper at the fire location.

Portable Water Tanks   Collapsible or inflatable temporary tanks for the storage of water that is dumped from tankers or tenders. Usually carried by the tender to set up a dump site.

Positive Pressure   A feature of SCBA providing a continuous supply of air, delivered by the regulator to the face piece, keeping toxic gases from entering. This pressure (1½ to 2 psi, depending on the manufacturer) is slightly above atmospheric pressure.

Post Indicator Valve (PIV)   A control valve that is mounted on a post case with a small window, reading either "OPEN" or "SHUT."

Post-Incident Thought Patterns   A phenomenon that describes an individual's inattentiveness following a significant incident. Post-incident thought patterns can lead to injuries or even death.

Preaction System   Similar to the dry pipe and deluge systems. The system has closed piping and heads with air under no or little pressure, but the water does not flow until signaled open from a separate fire detection system. The preaction valve then opens and allows water to flow through the system to the closed heads. When an individual head is heat activated, it opens and water attacks the fire. Usually used when water can cause a large dollar loss.

Prearrival Instructions   Self-help instructions intended to enhance the overall safety of the citizen until first responders arrive on the scene.

Pre-Incident Management   Advance planning of fire-fighting tactics and strategies or other emergency activities that can be anticipated to occur at a particular location. Often referred to as preplanning.

Pressure   The force, or weight, of a substance, usually water, measured over an area.

Pressure-Regulating Device   Designed to control the head pressure at the outlet of a standpipe system to prevent excessive nozzle pressures in hoselines.

Primary Hole   Ventilation term used to describe the first holes to be cut in a roof. They must be located as close to directly over the fire as possible to prevent laterally drawing the fire across unburned areas.

Protein Foam   Made from chemically broken down natural protein materials, such as animal blood, that have metallic salts added for foaming.

Prying Tools   The group of tools used to separate objects by means of a mechanical advantage.

Psychological Decon   The process performed when persons who have been involved in a situation think they have been contaminated and want to be decontaminated. Responders who have identified that the persons have *not* been contaminated should still consider what can be done to make them feel better.

Pulling Tools   The group of tools used to pull away materials.

Pulmonary Edema   Fluid filling the lungs causing death by drowning.

Pump Operator   A generic term to describe the person responsible for operating a fire apparatus pump. Other commonly used titles include motor pump operator, engineer, technician, chauffeur, and driver/operator.

Puncture   An injury caused by an object that has stabbed the body.

Purlins   A series of wood beams placed perpendicular to steel trusses to help support roof decking.

Pyrolysis   Decomposition or transformation of a compound caused by heat.

Quint   A combination fire service apparatus with components of both engine company and a truck company.

Rabbeted   A door stop that is cut (rabbeted) into the door frame. On metal door frames the stop is an integral part of the frame.

Radial Pulse   The pulse located in either wrist.

Radiological Dispersion Device (RDD)   An explosive device that spreads radioactive material throughout an area.

Rafter   A wood joist that is attached to a ridge board to help form a peak.

Rapid Intervention Crew (RIC)   See **Rapid Intervention Team.**

Rapid Intervention Team (RIT)   A company designated to search for and rescue trapped or lost firefighters. Depending on location, may also be called a FAST.

Rate of Spread   A ground cover fire's forward movement or spread speed. Usually expressed in chains or acres per hour.

RECEO   Acronym coined by Lloyd Layman standing for Rescue, Exposures, Confinement, Extinguishment, and Overhaul.

Recommended Exposure Limit (REL)   An exposure value established by NIOSH for a ten-hour day, forty-hour workweek. Similar to the PEL and TLV.

Reducers   Used to connect a larger hose to a smaller one.

Reel Coil   Memory that wire develops from having been placed on a wooden spool as it is being manufactured.

Regulations   Developed and issued by a governmental agency and have the weight of law.

Rehab   A shortened word meaning *rehabilitation.* Rehab typically consists of rest, medical evaluation, hydration, and nourishment.

Relief Valve   A device designed to vent pressure in a tank, so that the tank itself does not rupture due to an increase in pressure. In most cases these devices are spring loaded so that when the pressure decreases the valve shuts, keeping the chemical inside the tank.

Remote Shutoffs   Valves that can be used to shut off the flow of a chemical. The term *remote* is used to denote valves that are located away from the spill.

Reportable Quantity (RQ)   Both the EPA and DOT use the term. It is a quantity of chemicals that may require some type of action, such as reporting an inventory or reporting an accident involving a certain amount of the chemical.

Rescue   Those actions that firefighters perform at emergency scenes to remove victims from imminent danger or to extricate them if they are already entrapped.

Rescue Company   The unit designation of a group of firefighters assigned to perform specialized rescue work and/or tactics and functions such as forcible entry, search and rescue, ventilation, and so on.

Rescue Specialist   A firefighter with specialized training and experience in areas such as high angle rope rescue, confined space, trench, or structural collapse rescue.

Residential Sprinkler System   Smaller and more affordable version of a wet or dry pipe sprinkler system designed to control the level of fire involvement such that residents can escape.

Residual Pressure   The pressure in a system after water has begun flowing.

Respiratory Protection Programs   Management programs designed to ensure employee respiratory protection as required by OSHA 29 CFR 1910.134 and NFPA 1500.

Respiratory System   The system of the human body that exchanges oxygen and waste gases to and from the circulatory system.

Retard Chamber   Acts to prevent false alarms from a sudden pressure surge in the water supply by collecting a small volume of water before allowing a continued flow to the alarm device. The water from a surge is drained from a small hole in the bottom of the collection chamber.

Retention   The digging of a hole in which to collect a spill. Can be used to contain a running spill or collect a spill from the water.

Ricin   A biological toxin that can be used by a terrorist or other person attempting to kill or injure someone. It is the easiest terrorist agent to produce and one of the most common.

Ridge   The land running between mountain peaks or along a wide peak. A high area separating two drainages running parallel with them.

Ringdown Circuits   Telephone connection between two points. Going "off-hook" on one end of the circuit causes the telephone on the other end of the circuit to "ring" without having to dial a number.

Risk   The chance of injury, damage, or loss; hazard.

Risk-Based Response   An approach to responding to a chemical incident by categorizing a chemical into a fire, corrosive, or toxic risk. Use of a risk-based approach can assist the responder in making tactical, evacuation, and PPE decisions.

Risk/Benefit   An evaluation of the potential benefit that a task will accomplish in relationship to the hazards that will be faced while completing the task.

Risk Management   The process of minimizing the chance, degree, or probability of damage, loss, or injury.

River Bottom   Topographic feature where water runs from higher elevations to lower. Can be dry or wet depending on season or recent rains.

Rollover   A phenomenon where the burning of superheated gases from fire extends into the top areas of the compartment in the upper thermal layers.

Rope Hose Tool   About 6 feet (2 m) of ½-inch (13-mm) rope spliced into a loop with a large metal hook at one end and a 2-inch (50-mm) ring at the other. Used to tie in hose and ladders, carry hose, and perform many other tasks requiring a short piece of rope.

Round Turn   Formed by continuing the loop on around until the sections of the standing part on either side of the round turn are parallel to one another.

Run Card System   System of cards or other form of documentation that provides specific information on what apparatus and personnel respond to specific areas of a jurisdiction.

Running End   End of the rope that is not rigged or tied off.

Saddle A pass between two peaks that has a lower elevation than the peaks. Wind will pass through this area faster than over the peaks, so fire is drawn into this feature.

Safety Container A storage can that eliminates vapor release by using a self-closing lid. Also contains a flame arrestor in the dispenser opening.

Sea Containers Shipping boxes that were designed to be stacked on a ship, then placed onto a truck or railcar.

Search and Rescue Attempts by fire and emergency service personnel to coordinate and implement a search for a missing person and then effect a rescue.

Secondary Containment Any approved method that will prevent the runoff of spilled hazardous materials and confine it to the storage area.

Secondary Hole A ventilation hole that is opened only after the primary holes have been opened. It complements the primary holes.

Sectional View A vertical view of a structure as if it were cut in two pieces. Each piece is a cross section of the structure showing roof, wall, horizontal floor construction, and the location of stairs, balconies, and mezzanines.

Sector An area established and identified for a specific reason, typically because a hazard exists within the sector. The sectors are usually referred to as hot, warm, and cold sectors and provide an indication of the expected hazard in each sector. Sometimes referred to as a zone.

Self-Accelerating Decomposition Temperature (SADT) Temperature at which a material will ignite itself without an ignition source present. Can be compared to ignition temperature.

Self-Contained Breathing Apparatus (SCBA) A type of respiratory protection in which a self-contained air supply and related equipment are worn or attached to the user. Fire service SCBA is required to be of the positive pressure type.

Sensitizer A chemical that after repeated exposures may cause an allergic-type effect on some people.

Setting The finishing step, making sure that the knot is snug in all directions of pull.

Severance The cutting off of components (i.e., brake pedal, steering wheel) in a vehicle.

Shear A force that tends to tear a material by causing its molecules to slide past each other.

Sheetrock A trademark and another name for plasterboard.

Shelter in Place A form of isolation that provides a level of protection while leaving people in place, usually in their homes. People are usually sheltered in place when they may be placed in further danger by an evacuation.

Shock A serious condition caused by a problem or failure of the circulatory system that results in a decrease of oxygen and vital nutrients to the body's tissues. Also known as hypoperfusion.

Shock Load A load or impact being transferred to a rope suddenly and all at one time.

Shoring The use of timbers to support and/or strengthen weakened structural members (roofs, floors, walls, etc.) in order to avoid a secondary collapse during the rescue operation.

Short-Term Exposure Limit (STEL) A fifteen-minute exposure to a chemical followed by a one-hour break between exposures. Only allowed four times a day.

Shoulder Load Hose load designed to be carried on the shoulders of firefighters.

Shuttle Operation The cycle in which mobile water supply apparatus is dumped, moves to a fill site for refilling, and is returned to the dump site.

Siamese A device that connects two or more hoselines into one line with either a clapper valve or gate valve to prevent loss of water if only one line is connected.

Sick Building A term that is associated with indoor air quality. A building that has an air quality problem is referred to as a *sick building*. In a sick building, occupants become ill as a result of chemicals in and around the building.

Sick Building Chemical When a building is referred to as a *sick building*, certain chemicals exist within that cause health problems for the occupants. These chemicals are referred to as *sick building chemicals*.

Simple Beam A beam supported at the two points near its end.

Slab Door See **Flush** or **Slab Doors**.

Slot Loads Narrow section of a hose bed where hose is flat loaded in the slot.

Small Lines or **Small-Diameter Hose** Hose less than 2½ inches (63 mm) in diameter.

Soft Suction Hose Large-diameter woven hose used to connect a pumper to a hydrant. Also known as a soft sleeve.

Solid-Core Doors Doors made of solid material such as wood or having a core of solid material between face panels.

Solid Stream Nozzles Type of nozzle that delivers an unbroken or solid stream of water to the fire. Also called solid tip, straight bore, or smooth bore.

Solid A state of matter that describes materials that may exist in chunks, blocks, chips, crystals, powders, dusts, and other types. Ice is an example.

Solubility A liquid's ability to mix with another liquid.

Spalling Deterioration of concrete by the loss of surface material due to the expansion of moisture when exposed to heat.

Span of Control The ability of one individual to supervise a number of other people or units. The normal range is three to seven units or individuals, with the ideal being five.

Spanner Wrenches Used to tighten or loosen couplings. They may also be useful as a pry bar, door chock, gas valve control, and so on.

Speaking Trumpet Trumpet used by a foreman or crew boss to shout orders above the noise of fire-fighting activities.

Special Egress Control Device Door hardware that will release and unlock the door a maximum of fifteen seconds after it has been activated by pushing on the bar.

Specialist Level A level of training that provides for a specific type of training, such as railcar specialist; someone who has a higher level of training than a technician.

Specific Gravity Weight of a liquid in relation to water. Water is rated 1.

Specification (Spec) Plates All trucks and tanks have a specification plate that outlines the type of tank, capacity, construction, and testing information.

Spinal Immobilization The process of protecting patients against further injury by securing them to a backboard or other rigid device designed to minimize movement.

Sprain Injury to the ligaments that hold the body's joints together and allow them to move.

Sprinkler Systems Designed to automatically distribute water through sprinklers placed at set intervals on a system of piping, usually in the ceiling area, to extinguish or control the spread of fires.

Staging Part of the operations section where apparatus and personnel assigned to the incident are available for deployment within three minutes.

Stairwell An enclosed stairway attached to the side of a high-rise building or in the center core of same.

Standard of Care A legal term that means for every emergency medical incident, an emergency responder should treat the patient in the same manner as another emergency responder with the same training.

Standard Operating Procedure (SOP) Specific information and instruction on how a task or assignment is to be accomplished.

Standard Transportation Commodity Code (STCC) A number assigned to chemicals that travel by rail.

Standards Usually developed by consensus groups establishing a recommended practice or standard to follow.

Standing Part The part of a rope that is not used to tie off.

Standpipe Systems Piping systems that allow for the manual application of water in large buildings.

State Emergency Response Committee (SERC) A group that ensures that the state has adequate training and resources to respond to a chemical incident.

States of Matter Describes in what form matter exists, such as solids, liquids, or gases.

Static A rope having very little (less than 2 percent) elongation at normal safe working loads.

Static Pressure The pressure in the system with no hydrants or water flowing.

Staypoles The stabilizer poles attached to the sides of Bangor ladders that are used to assist in the raising of this type of ladder. Once raised, they are not used to support the extended ladder.

Steepness of Slope The degree of incline or vertical rise to a given piece of land.

Storz Couplings The most popular of the nonthreaded hose couplings.

Straight Stream A nozzle pattern that creates a hollow stream, similar in shape to the solid stream pattern, but the straight stream pattern must pass around the baffle of the nozzle. Newer fog nozzle designs, especially the automatic nozzles, only have this hollow effect from the tip on and, hence, create a solid stream with good reach and penetration abilities, some better than solid stream nozzles.

Strainers Placed over the end of a suction hose to prevent debris from being sucked into the pump. Some strainers have a float attached to keep them at or near the water's surface. A different style of strainer or screen is located on each intake of a pump.

Strategic Goals The overall plan developed and used to control an incident.

Stream Shape The arrangement or configuration of the water or other agent droplets as they leave the nozzle.

Stream Straighter A metal tube, commonly with metal vanes inside it, between a master stream appliance and its solid nozzle tip. The purpose is to reduce any turbulence in the stream, allowing it to flow straighter.

Strike Plate The metal piece attached to a door jamb into which the lock bolt slides. Also called a *strike* or *striker.*

Striking Tools The group of tools designed to deliver impact forces to break locks or drive another tool.

Sublimation The ability of a solid to go to the gas phase without being liquid.

Superficial Burns Burns affecting the outermost layer of skin, which typically cause redness of the skin, swelling, and pain.

Superfund Amendments and Reauthorization Act (SARA) A law that regulates a number of environmental issues, but is primarily for chemical inventory reporting by industry to the local community.

Supplied Air Respirator (SAR) A type of SCBA in which the self-contained air supply is remote from the user, and the air is supplied by means of air hoses.

Supply Hose or **Large-Diameter Hose (LDH)** Larger hose [3½ inches (90 mm) or bigger] used to move water from the water source to attack units. Common sizes are 4 and 5 inches (100 to 125 mm).

Surface-to-Mass Ratio Exposed exterior surface area of a material divided by its weight.

Suspension System The springs, shock absorbers, tires, and so on, of a vehicle.

Tactics The specific operations performed to satisfy the strategic goals for an incident.

Tactilely Using the sense of touch to feel for any differences or abnormality.

Tag/Guide Lines   Tag lines are ropes held and controlled by firefighters on the ground or lower elevations in order to keep items being hoisted from banging against or getting caught on the structure as they are being hoisted.

Tanker   The term given to aircraft capable of carrying and dropping water or fire retardant. Some departments still use the term to describe land-based water apparatus.

Target Hazard   An occupancy that has been determined to have a greater than average life hazard or complexity of firefighting operations. Such occupancies receive a high priority in the pre-incident management process and often a higher level of first-alarm response assignment.

TDD   Device that allows citizens to communicate with the telecommunicator through the use of a keyboard over telephone circuits instead of voice communications.

Teamwork   A number of persons working together in an effort to reach a common goal.

Technician Level   A high level of training that allows specific offensive activities to take place, to stop or handle a chemical incident.

Telecommunicator   Individual whose primary responsibility is to receive emergency requests from citizens, evaluate the need for a response, and ultimately sound the alarm that sends first responders to the scene of an emergency.

Tempered Glass   Plate glass that has been heat treated to increase its strength.

Tender   The abbreviated term for *water tender.* A water tender is defined as a land-based mobile water supply apparatus. Some departments still use the term *tender* to describe a hose-carrying support apparatus.

Tensile Strength   Breaking strength of a rope when a load is applied along the direction of the length, generally measured in pounds per square inch.

Tension   A force that pulls materials apart.

Terra Cotta   Tiles composed of clay and sand that are kiln fired. May be structural or decorative.

Terrorism   Acts of violence that are arbitrarily committed against lives or property and intended to create fear and anxiety.

Tether Line   A rope that is held by a team on shore during a water rescue to be used to haul the rescuer and victim back to shore.

Thermal Burns   Burns caused by heat or hot objects.

Thermal Layering   The stratification of gases produced by fire into layers based on their temperature.

Thermal Level   A layer of air that is of the same approximate temperature.

Thermal Plume   A column of heat rising from a heat source. A fully formed plume will resemble a mushroom as the upper level of the heat plume cools, stratifies, and begins to drop outside the rising column.

Thermal Protective Performance (TPP)   A rating level, expressed in seconds, used to characterize the protective qualities of a PPE component before serious injury is experienced by the wearer.

Threshold Limit Value (TLV)   An exposure value that is similar to the PEL, but is issued by the ACGIH. It is based on an eight-hour day.

Through-the-Lock Method   A method of forcible entry in which the lock cylinder is removed by unscrewing or pulling and the internal lock mechanism is operated to open a door. Also, the family of tools used to perform this operation.

Tidal Changes   The rising and falling of the surface water levels due to the gravitational effects between the Earth and the moon. In some areas, these changes are insignificant but in others there is more than 40 feet of difference between high and low tide.

Tip Arc   The path that a ladder's tip will take while being raised.

Torsion Load   A load parallel to the cross section of the supporting member that does not pass through the long axis. A torsion load tries to "twist" a structural element.

Total Flooding System   Used to protect an entire area, room, or building by discharging an extinguishing agent that completely fills or floods the area with the extinguishing agent to smother or cool the fire or break the chain reaction.

Tote   A large tank usually 250 to 500 gallons, constructed to be transported to a facility and dropped for use.

Tower Ladder   An apparatus with a telescopic boom that has a platform on the end of the boom or ladder. It can be extended or retracted and rotated like an aerial ladder.

Toxicity   Poisonous level of a substance.

Toxins   Disease-causing materials that are extremely toxic and in some cases more toxic than other warfare agents such as nerve agents.

TRACEM   An acronym for the types of hazards that exist at a chemical incident: thermal, radiation, asphyxiation, chemical, etiological, and mechanical.

Triage   A quick and systematic method of identifying which patients are in serious condition and which patients are not, so that the more seriously injured patients can be treated first.

Triple Combination Engine Company   Fire apparatus that can carry water, pump water, and carry hose and equipment.

Truck Company   The unit designation of a group of firefighters assigned to perform tactics and functions such as forcible entry, search and rescue, ventilation, and so on.

Truss   A rigid framework using the triangle as its basic shape.

Tunnel Vision The focus of attention on a particular problem without proper regard for possible consequences or alternative approaches.

Tunneling The digging and debris removal accompanied by appropriate shoring to safely move through or under a pile of debris at a structural collapse incident.

Turntable The rotating platform of a ladder that affords an elevating ladder device the ability to turn to any target from a fixed position.

Two In/Two Out The procedure of having a crew standing by completely prepared to immediately enter a structure to rescue the interior crew should a problem develop.

Type A Reporting System System in which an alarm from a fire alarm box is received and retransmitted to fire stations either manually or automatically.

Type B Reporting System System in which an alarm from a fire alarm box is automatically transmitted to fire stations and, if used, to outside alerting devices.

Type I, Fire-Resistive Construction Type in which the structural members, including walls, columns, beams, girders, trusses, arches, floors, and roofs, are of approved noncombustible or limited combustible materials with sufficient fire-resistive rating to withstand the effects of fire and prevent its spread from story to story.

Type II, Noncombustible Construction Type not qualifying as Type I construction, in which the structural members, including walls, columns, beams, girders, trusses, arches, floors, and roofs, are of approved noncombustible or limited combustible materials with sufficient fire-resistive rating to withstand the effects of fire and prevent its spread from story to story.

Type III, Ordinary Construction Type in which the exterior walls and structural members that are portions of exterior walls are of approved noncombustible or limited combustible materials, and interior structural members, including walls, columns, beams, girders, trusses, arches, floors, and roofs, are entirely or partially of wood of smaller dimension than required for Type IV construction or of approved noncombustible or limited combustible materials.

Type IV, Heavy Timber Construction Type in which exterior and interior walls and structural members that are portions of such walls are of approved noncombustible or limited combustible materials. Other interior structural members, including columns, beams, girders, trusses, arches, floors, and roofs, shall be of solid or laminated wood without concealed spaces.

Type V, Wood Frame Construction Type in which the exterior walls, bearing walls, columns, beams, girders, trusses, arches, floors, and roofs are entirely or partially of wood or other approved combustible material smaller than the material required for Type IV construction.

Underground Storage Tank (UST) Tank that is buried under the ground. The most common are gasoline and other fuel tanks.

Undesigned Load A load not planned for or anticipated.

Unified Command The structure used to manage an incident involving multiple response agencies or when multiple jurisdictions have responsibility for control of an incident.

Unity of Command One designated leader or officer to command an incident.

Upper Explosive Limit (UEL) The upper part of the flammable range. Above the UEL, fire or an explosion cannot occur because there is too much fuel and not enough oxygen.

Utility Rope Rope used for utility purposes only. Some of the tasks utility ropes are used for in most every fire department are hoisting tools and equipment, cordoning off areas, and stabilizing objects. Also used as ladder halyards.

Vacuum (Negative) Pressure The measurement of the pressure less than atmospheric pressure, which is usually read in inches of mercury (in. Hg or mm Hg) on a compound gauge.

Vapor Density Weight of a gas in relation to air. Air is rated 1.

Vapor Dispersion The intentional movement of vapors to another area, usually by the use of master streams or hoselines.

Vapor Pressure The amount of force that is pushing vapors from a liquid. The higher the force the more vapors (gas) being put into the air.

Vapor Suppression Ability to contain or control the production of fuel vapors.

Vaporization The process in which liquids are converted to a gas or vapor.

Variable, Adjustable, or Selectable Gallonage Nozzle Nozzle that allows the nozzleperson to select the flow, with usually two or three choices, and the pattern.

Veins The blood vessels, or tubes, within the body that carry blood lacking oxygen and nutrients back to the heart.

Velocity Pressure The forward pressure of water as it leaves an opening.

Veneer A covering or facing, not a load-bearing wall, usually with brick or stone.

Venous Bleeding Bleeding from a vein.

Venturi Principle A process that creates a low-pressure area in the induction chamber of the eductor and allows the foam concentrate to be drawn into and mix with the water stream.

Vertical Ventilation Channeled pathway for fire ventilation via vertical openings.

Vesicants A group of chemical agents that cause blistering and irritation of the skin. Commonly referred to as blister agents.

Vicarious Experience   A shared experience by imagined participation in another's experience.

Visqueen   A trade name for black plastic. It can be used very effectively in salvage and overhaul operations.

Voice Inflection   Change of tone or pitch of voice.

Voids   Spaces within a collapsed area that are open and may be an area where someone could survive a building collapse.

Wall Indicator Valve (WIV)   A control valve that is mounted on a wall in a metal case with a small window, reading either "OPEN" or "SHUT."

Watch Dogs   Trained dogs that will bark and create a commotion, but will not attack.

Water Columning   A condition in a dry pipe sprinkler system in which the weight of the water column in the riser prevents the operation of the dry pipe valve.

Water Curtain Nozzle   Designed to spray water to protect exposures against heat by wetting the exposure's surface.

Water Hammer   A sudden surge of pressure created by the quick opening or closing of valves in a water system. The surge is capable of damaging piping and valves.

Water Table   The level of groundwater under the surface.

Water Tender   The term given to land-based water supply apparatus.

Water Thief   A variation of the wye that has one inlet and one outlet of the same size plus two smaller outlets with all of the outlets being gated. The standard water thief usually has a 2½-inch (65-mm) inlet with one 2½-inch (65-mm) and two 1½-inch (38-mm) outlets.

Waybill   A term that may be used in conjunction with consist, but is a description of what is on a specific railcar.

Weapon of Mass Destruction (WMD)   A term that is used to describe explosive, chemical, biological, and radiological weapons used for terrorism and mass destruction.

Web   The vertical portion of a truss or I beam that connects the top chord with the bottom chord.

Web Gear   The term given to a whole host of personal items carried on a belt/harness arrangement worn by wildland firefighters. Items include water bottles, a fire shelter, radio, and day sack.

Webbing   Nylon strapping, available in tubular and flat construction methods.

Webbing Sling   Approximately 12 to 15 feet of rescue webbing tied end to end, forming a continuous loop.

Western or **Platform Framing**   A style of wood frame construction in which each story is built on a platform, providing fire stopping at each level.

Wet Chemicals   Extinguishing agents that are water-based solutions of potassium carbonate–based chemicals, potassium acetate–based chemicals, or potassium citrate–based chemicals, or a combination.

Wet Pipe Sprinkler System   Has automatic sprinklers attached to pipes with water under pressure all the time.

Wheatstone Bridge Sensor   A type of combustible gas sensor that uses a heated coil of wire to determine the presence of flammable gases.

Wire Glass   Glass with a wire mesh embedded between two or more layers to give increased fire resistance.

Work Hardening   A phrase given to the effort and physical training designed to prepare an individual to better perform the physical tasks that are expected of the individual. Work hardening is key in preventing injuries resulting from typical fire-fighting tasks.

Working End   The end of the rope that is utilized to secure/tie off the rope.

Working Length   The length of the ladder that spans the distance from the ground to the point of contact with the structure. This does not include any distance the ladder might go beyond the point of contact as would be the case when the tip extends beyond the roof.

Wye   A device that divides one hoseline into two or more. The wye lines may be the same size or smaller size and the wye may or may not have gate control valves to control the water flow.

Zone   An area established and identified for a specific reason, typically because a hazard exists within the zone. The zones are usually referred to as hot, warm, and cold zones and provide an indication of the expected hazard in each zone. Sometimes referred to as a sector.

Zoning   A term given to the establishment of specific hazard zones; that is, hot zone, warm zone, cold zone. Also collapse zones.